The Affordable Immortal:

Maybe You *Can* Beat Death and Taxes!

Rudi Hoffman
World's Leading Cryonics Insuror

© 2018 Rudi Hoffman, all rights reserved.

No part of this book may be reproduced, stored in a retrieval system or transmitted by any means without the sole consent and written permission of the author.

First published in April 2018
Edited by Sean Donovan

ISBN-13: 978-1986985369
ISBN-10: 1986985369

Printed in the United States of America

Contents

Introduction: Welcome to the Future 1
   1. A Life or Death Decision: WWYD? 7
   2. A Time Machine to the Future 21
   3. What's the Fare on this Time Machine?......... 31
   4. Jerry's Epic Cryonics Odyssey 53
   5. A Non-Boring Chapter on Life Insurance 63
   6. Is Your Life Better than Grandma's?............... 81
   7. Watercooler Wisdom 101
   8. The Merkle Matrix 117
   9. Objections Overruled 129
   10. Happy Wife, Happy Life 141
   11. Any Excuse for a Vacation! 149
   12. Mecca for Nerds 153

Appendices... 171
Contacts and References 179
Acknowledgements 187
About the Author 189

To YOU, the reader: Benjamin Franklin once said, "In this world nothing can be said to be certain, except death and taxes."
Well, times have changed.
Take advantage.

"I wish it were possible, from this instance, to invent a method of embalming drowned persons in such a manner that they may be recalled to life at any period, however distant; for having a very ardent desire to see and observe the state of America a hundred years hence, I should prefer to any ordinary death the being immersed in a cask of Madeira wine with a few friends till that time, to be then recalled to life by the solar warmth of my dear country!"

~ *Benjamin Franklin, in a 1773 letter.*

Introduction: Welcome to the Future

Why This Book Was Written
My mission in this book is two-fold. First, to cover some of the ideological assumptions which underlie cryonics as an emerging technology.

Second, to propose that cryonics is financially feasible for you, if you are fairly healthy and have some reasonable but not necessarily extreme financial resources. Whether reanimation is possible or not becomes a moot point to you if cryonics is only available to the affluent or would require great sacrifice on the part of your loved ones.

What do I hope you get out of this book? Here are five ideas I would like you to consider. Figuring out an interesting way to convey these foundational memes has kept me wide awake many nights. You will be the judge of whether I succeed in framing these themes in a cogent and compelling way.

Five Takeaways
1. Cryonics is a legitimate, though currently unproven, medical intervention.
2. Assuming the above, you may want to be in the cryonics "experimental group" and not in the "control group."
3. This choice may be affordable for you through the leverage of life insurance.
4. If cryonics does indeed work and you are revived, it will probably be in a really spectacular and fun future.
5. There are resources and people to help you in your research and decision making. I am one of those people.

Yes, it just may be possible for you to beat death and taxes! This book is written to explain why that sentence is not as unlikely (crazy?) as it may seem. I acknowledge this is a mind-stretching claim, and I welcome your skepticism.

This book will explain how and why most individuals might reasonably incorporate the amount of money required for cryonic suspension into their budget. This is generally accomplished through the financial "leverage" of life insurance, where a relatively small amount of premium paid to an insurance company blossoms to an enormous amount of money upon pronouncement of "death."

The cryonics procedure is not actually carried out until there is a medical professional's pronouncement of "death." At that point, when an individual is pronounced "dead" by current *legal* (not necessarily medical) standards, any life insurance policies on the individual are fully collectible and will be paid by the life insurance company.

What this means, in practical terms, is that nearly everyone reading these words, and I do include *you*, dear reader, now has the financial ability to afford this potentially life-preserving technology.

This Is a Huge Deal!
If this does not astonish you, if you do not sense the life-changing and society-transforming power of this simple idea, then I have yet to do my job and you should perhaps reconsider the implications of these technologies. By the time we finish the journey of this book together, my hope is that you will deeply consider the life-enhancing possibilities that are now available to millions of irreplaceable and unique individuals.

But Will Cryonics Work?
Whether or not you consider cryonic suspension as a rational or reasonable gamble is your decision, based on research that you have perhaps already done and will continue to do. Whether cryonics will work and whether the cryogenically suspended

individual can be reanimated with memories intact and able to function and enjoy life is an open-ended question. While this is perhaps among the more important questions of our age, and an especially vital question for you to make regarding your personal cryonics choices, this book will not focus primarily on the technical feasibility of cryonics.

Disclaimer
It is considered good form, especially in scientific literature, to disclose bias. With this in mind, the reader should be warned that my bias is that cryonic suspension is an enormously reasonable, rational, affordable, and scientific choice. You, as a perceptive reader, have probably figured this out already.

Isn't cryonic suspension just science fiction?

The answer to this question is that cryonic suspension is a reality, although future revival remains an open question. We'll go into explicit detail about the current realities of this fascinating endeavor as we proceed.

A major premise of cryonics is the expectation that future science and technological advances will, at some point in time, be able to restore an individual with their memories and sense of self intact.

There is a consciousness in the 1.5-kilogram universe sitting between your ears as you read these words, remarkably similar to the awareness I feel as I craft these same words. I intend to create verbiage that *may* change your worldview in a profound way, a bridge from my mind to yours. I want to share some ideas that may preserve your consciousness, whatever you consider to be at the very heart of "you."

Let's get started. This will be fun. And, if you approach the ideas in this book with a reasonably skeptical but open mind, these ideas may impact your life dramatically in a positive direction. Would you choose life in Chapter One's scenario?

What would you do?

Chapter 1
A Life or Death Decision: WWYD?

Ponder this very real and gritty metaphorical story… What Would You Do?

You never thought it would happen to you. "There must be some mistake in the diagnosis" you tell your doctor and your wife. "I have never smoked, I eat a healthy and mostly vegetarian diet, I work out a few times a week, and I am training for that half-marathon coming up in two months. Doctor, are you sure you got this right?"

Your attention is riveted on the rail-thin oncologist in his starched lab coat. You glance over at your wife to see how she is taking this, and notice her brave face, but you can also see the detail of how her eyes seem especially bright, as the tears she will not let fall well up in her eyes. You have been waiting with her in this tiny room for the doctor to come in and discuss the results of the PET scan you had last week.

The doctor still seems too young, but you have heard he is the best available. And you have had some deep, even philosophical, conversations with him. Could it just be three weeks ago that you and your wife were making plans for your long-awaited and much needed vacation? After many years of postponed vacations and planned trips, this trip to Hawaii was going to be the trip of a lifetime; a way to make up for the many late nights and missed dinners that were just a part of doing "whatever it takes" to make your business work. She understood, of course, but this trip was the one you both had promised each other would not be postponed - no matter what.

You consider whether it would be wildly inappropriate to grin at your wife sardonically and say, "Welcome to 'no matter what.'"

You ponder the best way to handle this. Conflicting demands vie for attention in your thought process. You want to set the right positioning and posture with this doctor, although a part of you wonders why this could possibly matter. Another voice inside your head wants to impress your wife with the wisdom and maturity you are showing in the face of what seems to be a very serious cancer diagnosis. Yet a third part of you wants a way out of this, to find a way to minimize how this is going to be impacting your life and plans. You are already negotiating with death, surprised that you feel more sadness over the lost Hawaiian vacation than anything else, and you speculate it is because the "anything else" does not feel real as of now. Yet a fourth voice in your head is rating the doctor on his professionalism, judging him on a one to ten scale as to how he delivers bad news.

The doctor may be too young, but he has obviously had many awkward conversations like this before. Much to your surprise, he is forthright and unvarnished. His directness is startling.

The Diagnosis
"I am afraid I have some bad news. You have an inoperable brain tumor we call a glioblastoma. There is simply no way to operate on it. Traditional chemotherapy and radiation are not effective options."

You simply cannot believe what you are hearing. Surely this is not happening! You clear your throat, so you can speak with a level and rational voice. "Doctor, what is the prognosis? What do we expect to happen, and exactly what are we going to do about it?"

As you end this question you look deep in this doctor's unlined face. You burn a hole in his left eye with your steady gaze. Surely, if the universe awards points for handling bad news in a rational way, you will get some for the steadiness you are showing, the fixity of your gaze into the doctor's surprisingly deep blue eyes.

Some part of you notes this is somewhat silly and irrational in itself. You struggle to maintain eye contact and note an unusual pressure in your own eyes. It is an odd sensation, and you realize this is the feeling of tears welling up as you search your doctor's face for something resembling hope.

He maintains his eye contact with you, but as he does, he also says words you never expected to hear. "There is simply no operating cure with this particular type of glioblastoma. It is deep in your brain stem. You need to get your affairs in order. I have consulted with the radiologist and my colleagues and our best guess is that you have between two to six months left to live. To be honest, with this type of mass and at stage three, it is probably closer to the two months. You will start to notice balance problems at first, then your legs, and later your arms, will become uncontrollable, followed by loss of function in your core organs."

Again, the feeling of unreality overcomes you. You feel like a character in a television drama, hearing some lines, and then you realize you don't know what your lines should be. You wait for the director to yell "Cut!" before realizing that this is you, not a character in some fiction show. This moment you are experiencing is about as real as life gets. You have just been given a death sentence.

The small room seems even smaller now. The part of you that always loved dramatic dialogue notes, with appreciation, that no outside noises or distractions from the hall interrupt this dramatic moment being shared between three human beings. In the sudden stillness, you hear your wife stifling some quiet sobs, and you want to comfort her.

Most of all, you want to fix this problem. You have taken an engineering approach to many problems in life, and there surely must be some way to fix this massive dilemma. Your business heroes and mentors became your role models because they took on difficult or impossible challenges. Can't this doctor see that you are special, that you don't deserve to have your life cut short by some incredibly stupid lump of tissue that doesn't know enough to stop growing?

"Listen, Doctor," you are pleased to note how calm and compelling your voice sounds in the confines of this room. "I believe in technology, human rationality, engineering, and enlightenment values. We are not in the dark ages. There simply must be something we can do!"

The quiet intensity of your question reaches beyond the professional facade of this young man.

You feel the deep connection that is so rare between humans. You know he has spent thousands of hours of his life so that he can help people like you in this exact situation. And you sense his deep compassion, which goes well beyond his professional obligation. You recall earlier conversations with him and know that he shares much of your worldview. You share a passion for what science and technology have done for humanity and almost an equally strong revulsion for superstitions, which make for bad decision making even among supposedly modern people.

"Well," the doctor hesitates, subtly shifting in his chair. He is now even closer, confirming your earlier belief that he genuinely likes you personally, and that delivering this news is breaking his heart. Your wife and you, seated in the plastic chairs in this tiny alcove, could reach out and touch each other. The physical intimacy is somewhat uncomfortable. You can see your doctor's stubble on the left side of his deeply cleft chin. The part of you that plays director in your head notes with approval that this guy could be from central casting to play the part of the celebrated oncologist, series superstar, and heartthrob of attractive nurses. But this is not the vicarious experience of movies and television. This is real. It is happening to you, in real time.

"Listen, Doc!" You realize the outwardly calm demeanor you have been projecting is not going to be maintained. Maybe it is time to show some intensity, to use some colloquial and decidedly unprofessional language to connect on a more personal level. "Goddammit, Doc! I do not intend to die in a few months! If there is anything… anything that can be done, I want to know about it!" Your voice has become suddenly very loud. Although you are almost within touching distance of these two people, you realize you are shouting at maximum volume.

You look over at your wife to see how she is handling your outburst. She is staring at you with an expression that is hard to read. Anger? Love? Grief? Embarrassment? Respect? What does she want from you?

You deeply want her to feel that you are handling this news without resorting to the kind of religious platitudes that you both have agreed are not helpful to the human condition. "It is all in the hands of God," is one such phrase you both had agreed on in earlier discussions, which basically pleads ignorance and abdicates personal accountability. If this cancer was a type that was curable with treatment, which has a good chance of working, would we "Leave it up to God?" Of course, we would not! Intelligent humans, living in the 21st century, simply do not do that kind of thing. So, what do intelligent human beings do when faced with imminent death?

"Well," he continues, "There is one chance. There is an experimental treatment for your condition; however, it is still in medical trials. It is not FDA approved and we don't know what the odds are of this treatment working. It is called gene therapy. You should also know that, because it is not FDA approved, your insurance company will not pay for it. It is quite expensive. While we have lots of proof of concepts that this should work, and we have clinical and lab experiments on animals which look promising, we cannot guarantee your results or even give you accurate odds that this will work. We may be able to get you into an experimental trial, in which case most of your costs would be covered; however, you should have no illusions that we can guarantee the outcome. But, we *can* guarantee that you are most likely going to die if you don't sign up for this clinical trial."

"Now we are talking! Now we can talk about making a smart decision!" you say with some relief. You had, in earlier discussions, asked your doctor to be straight with you, but you are still surprised by his breathtaking directness. At the same time, your training on rational decision making enables you to realize that making important decisions in the heat of the moment is often unwise. Is it now unwise to decide on an experimental therapy right now, in this moment, and to perhaps spend your life savings as well as what would be left to your wife and family, on an experimental treatment?

Your training in rational decision making drilled you on the many cognitive biases that even well-educated and smart people repeatedly demonstrate. Availability bias, you remind yourself, is the fact that we simply don't know what it is we don't know.

In an attempt to get back in touch with your executive decision-making forebrain, you run some of the other well-known cognitive glitches that we humans are prone to, replaying the parts of the list you can remember through your mind. *Availability bias, recency bias, confirmation bias, ignoring relevant information, neglect of probability, anchoring, self-serving bias, fundamental attribution error, belief bias, framing bias, and hindsight bias* all reel through your mind. While you know this list is incomplete (in itself, you reflect, a type of *availability bias*), you know that we, as humans, have developed a solid repertoire of ways to fool ourselves and others.

Your doctor has just told you that you are going to die. In essence, you are going to drown in a sea of ignorance because modern medicine has simply not figured out how to fix you yet. And he has thrown you a life preserver, but this life preserver does not have a stamp of approval from regulatory agencies and mainstream medicine.

You wonder, "Am I being stupid, pathetic, and desperate to consider a non-FDA approved treatment? I have always been harsh in my criticisms of those who seem to reject science in favor of new age type modalities. I have had such a smug superiority to them, feeling that they simply lack the empiricism and scientific training that I have tried my best to learn over many years of study. They believe something is true because they simply want it to be true. How embarrassing is that? Could we ask for any clearer example of wishful and magical thinking, the opposite of scientific decision making? Would my considering this experimental treatment put me in the same boat as new age crystal wavers and tarot card readers?"

Once again, life has given you an opportunity to regret your earlier arrogance. It turns out that making a real-world decision, with massive personal impact, in the face of unknown variables and even unknown unknowns, is perhaps easier to observe in a movie theatre than in real life - especially when it is your own life, or that of a loved one. The tendencies we all have, which may or may not be well informed by our rational minds, are certainly evidenced when the stakes are so high, like this life or death decision.

You ask yourself, "Should I take a chance to live, or should I do nothing and die?"

After a moment, you recognize that your decision is not one of science versus pseudoscience but is instead one of current and proven science vs. proto-science, an early and unproven, but reasonable and logical extension of current scientific trends.

The experimental treatment you are considering is a scientific and technological intervention, among the most medically advanced interventions possible.

What Do You Do Now?
In this scenario, which I hope has enough details to emotionally resonate, we see the human condition writ small; a single individual facing the fact that his life will end. The difference between him and you, as you read this sentence, is just a matter of time. Our hero had about two months. You have a little longer. Maybe.

On the scale of biological time or geological time, the 20 or 50 or 70 years you may have before you die is a mere flicker, a span of time so small that it can't even be seen at these scales.

One of the cognitive biases we all share is that it is difficult for us, even as reflective and deeply philosophical beings, to consider our own non-existence.

So, given a choice for an experimental treatment or certain death, which would you choose? Please be assured that you *will* be making a choice. By doing nothing and letting nature "take its course," you will almost certainly die. The difference between you and the scenario you just read is only a matter of time.

On a very personal note from the author to you, as I write these words in April of 2018, it is almost two years to the day that I woke up with a lump on my leg that turned out to be stage-2 lymphoma throughout my body and spleen. The parts of the above scenario that seem the most visceral are directly from my very recent life experience. I am still here writing these words, and you are reading them, as a result of successful hyper-current medical interventions.

Could it be that there may be an experimental treatment which could save you? Is it possible you could be fortunate enough to live in an age when science and technology may give you a chance to live and not die? How important would this be if it were true? Let's agree to simply keep an open mind and review some facts together in the journey of this book.

Chapter 2
A Time Machine to the Future

So, just what IS cryonics? Could cryonics be *your* time machine to the future?

This chapter is intended to provide basic information about cryonics to help you answer those questions. I hope that the level of explanation is appropriate for a well-informed individual who is curious about the science of cryonics as well as about the practical aspects of personally becoming involved with cryonics.

The purpose of this chapter is not to focus on the science or technological feasibility of cryonics; nor is it to provide the history of this intriguing field of endeavor, although that would be a different and fascinating chapter or even another book. Instead, it is a helpful overview of this fascinating emerging science.

Cryonics, as defined by Google:
cry·on·ics - (krī'äniks)
  noun

      The practice or technique of deep-freezing the bodies of people who have just died, in the hope that scientific advances may allow them to be revived in the future.

However, I prefer an even more straightforward definition, so I propose:
> Cryonics is an experimental procedure that preserves a human being using the best available technology for the purpose of saving his/her life.

Cryonics as a Medical Intervention
Here is a cartoon drawing with stick figures portraying the state of the human condition:

This cartoon is from a brilliant research publication on cryonics by Tim Urban. The simplicity of the figure is intentional, to show that you and everyone you know is going over the cliff of death. What if there were a bridge, or a ladder, to get to that far side? Would you take it, or would you follow the mainstream of humanity over the edge?

While conceptually simple, the actual practice of cryonics is medically and technologically complex. Like other medical procedures such as heart transplants, the idea is simple in concept, but difficult in execution.

Cryonics has further analogies to heart transplants, in that even the very idea has been controversial. As these words are written in 2018, the distinction between heart transplants and cryonics is significant. Heart transplants, which at one time were controversial and remain costly[1], are now considered mainstream medicine. Former United States Vice President Dick Cheney had a heart transplant at age 71. Costing from $500,000 to a million dollars, with no guarantee of success, heart transplants and other organ transplants are

---

[1] Should Violent Felons Receive Organ Transplants?

Imagine watching a loved one die for lack of a heart, then reading in the paper the story ... Markkula Center of Applied Ethics ... The cost of the transplant and years of post-operative care may reach $1 million, courtesy of California taxpayers.

www.scu.edu/ethics/publications/submitted/Perry/transplant.html

increasingly bringing us into the era of "science fiction" medical practice[2].

In contrast to organ transplants, as of this writing the actual science and technology of cryogenically preserving humans and higher animals remains in its infancy and unproven. While the cryogenic preservation of sperm, eggs, embryos, and some tissues is no longer even considered remarkable, the technological issues involved with preserving and reviving animals with millions or even trillions of cells remain significant.

There *are*, however, numerous "proof of concept" ideas related to the practice of cryonics. These are more accurately described not just as "proof of concept" but are instead actual current level science. These include:
1. Cryogenic storage and viability of frozen sperm, eggs, and embryos[3]

---

[2]Organ Donation Ethics - Southern Methodist University

Jan 8, 2006 ... The estimated costs for a heart transplant during the first ... according to Tom Mayo, a Southern Methodist University bioethicist and one of the ...
smu.edu/newsinfo/excerpts/cardiac-donation-ethics.html

[3]Experts: Egg freezing no longer 'experimental' - CNN.com

Oct 19, 2012 ... Egg freezing should no longer be considered "experimental," ... embryo implantation rates and pregnancy rates of fresh eggs versus eggs that ...
www.cnn.com/2012/10/19/health/egg-freezing

2. Surgical operations performed on humans who are cooled to reduce damage[4]
3. People who survive with their brain functions intact after "drowning" or being stuck in snow drifts[5]
4. Animals who go into a state of suspended animation and are later revived with complete functionality intact[6] [7]

---

[4]Patients to be frozen into state of suspended animation for surgery ...
Sep 26, 2010 ... The technique helps to reduce the damage done to the brain and other ... The normal human body temperature is 37 degrees C and usually ...
www.telegraph.co.uk/health/healthnews/8024991/Patients-to-be-frozen-into-state-of-suspended-animatio...

[5]Near drowning: MedlinePlus Medical Encyclopedia

It may be possible to revive a drowning person even after a long period under water, especially if the person is young and was in very cold water.
.www.nlm.nih.gov/medlineplus/ency/article/000046.htm

[6]Roth Lab - Labs - Fred Hutchinson Cancer Research Center

Our work in suspended animation derives from the fact that many animals exhibit what we call "metabolic flexibility," the ability to dial down their respiration and ...

labs.fhcrc.org/roth/

[7]Mark Roth on Mice and Men and Suspended Animation - Wired

5. The technological challenges that remain to be solved for "cryonics" with large animals like humans are significant. However, it may be important to realize that no responsible scientists assert that there are any laws of physics that would need to be violated in order for cryonics to "work". Moreover, there are many responsible and mainstream scientists who have endorsed the idea, at least on a conceptual basis.[8]

To be fair and even handed, it should be observed that cryonics in 2018 is still far from a "mainstream" idea. No less an authority on the possibilities of science and technology than Michio Kaku has an internet video asserting his opinion on "Why Cryogenics is Bogus."[9]

Feb 11, 2010 ... In this, and other tests of the product on animals, benefits were achieved, Roth says, without putting the animals into suspended animation.

www.wired.com/business/2010/02/mark-roth-on-mice-and-men/

[8] Scientists' Open Letter on Cryonics

Cryonics is a legitimate science-based endeavor that seeks to preserve ... of this letter does not imply endorsement of any particular cryonics organization or its ..

www.imminst.org/cryonics_letter/

[9] Why Cryogenics Is Bogus | Michio Kaku | Big Think

Regrettably, Michio Kaku is still misinformed by the two "GREAT MYTHS" of cryonics, based on the content of the above referenced video. These are:
Myth 1: Current cryonics protocols do nothing to reduce freezing damage and
Myth 2: Cryonics is only for very wealthy people.

Kaku is disappointingly poorly informed on both of these ideas, as even a cursory bit of research shows both these myths to be unambiguously false.

It may be helpful at this point to be reminded that many, if not most, medical interventions have gone from being considered "freakish" to "of course."

Anesthesia, defibrillation, in vitro fertilization, and organ transplantation, for example, were all decried at some point as immoral. Some church leaders and so-called "ethicists" expressed indignation and shock about "man playing God."

However, now all of these medical interventions are considered to be reasonable options, indeed a standard of care. Interventions that were formerly

---

Jan 26, 2011 ... When you freeze human tissue, it may appear to be preserved superficially, but the ice crystals that form create massive cell damage, causing ...

bigthink.com/videos/why-cryogenics-is-bogus

controversial are now almost a right to the patient and indeed an obligation of modern medicine.

Is it possible that the medical bio stasis of cryonics may follow a similar path?

Cryonics as Currently Practiced

Cryonics is far from a new idea. Robert Ettinger, in his seminal book published in 1964, *The Prospect of Immortality,* provided the basic framework of "freeze, wait, reanimate." He later founded the Cryonics Institute near Detroit, Michigan, where he currently resides while waiting resuscitation.[10]

There are now perhaps half a dozen organizations throughout the world that are engaging in the practice of cryonics. The best known and most established are the two main cryonics organizations we will focus on in this book, Ettinger's Cryonics Institute and the Alcor Life Extension Foundation.

---

[10] Robert Ettinger, founder of the cryonics movement, dies at 92 ...

July 24, 2011 ... Robert C. W. Ettinger, a physics teacher and science fiction writer who believed death is only for the unprepared and unimaginative, died July ...

articles.washingtonpost.com/2011-07-24/local/35237048_1_cryonics-institute-cryonics-movement-deep-fr.

Both of these organizations have been quietly and solidly in place for decades. Alcor was founded in 1972, the Cryonics Institute a few years later.

While unproven, cryonics is an option and a personal choice that several thousand people have been making for themselves since the inception of the idea.

As of this writing, the number of people actually fully signed and funded for cryonics is surprisingly small. Membership at Alcor Life Extension Foundation and the Cryonics Institute, along with the American Cryonics Society and a few developing cryonics facilities throughout the world, total a membership less than 3000.[11] [12] Still, it seems that this is a "risk/reward" determination best left to reasonably well-informed individuals.

---

[11]Member Statistics Details - Cryonics Institute

Cryonics Institute information concerning statistics details.

www.cryonics.org/statistics_details.html

[12]Alcor Life Extension Foundation - Wikipedia, the free encyclopedia

The largest cryonics organization today, in terms of membership, was established as a nonprofit organization by Fred and Linda Chamberlain in California in ...

en.wikipedia.org/wiki/Alcor_Life_Extension_Foundation

Chapter 3
What's the Fare on
This Time Machine?

So, if cryonics is a potential time machine, what are the costs involved to buy transport? And is this technology affordable for you, the brain that is reading these words?

As noted in the last chapter, there are two main cryonics organizations doing business in the United States today (2018), Alcor Life Extension Foundation, hereafter referred to as Alcor, and the Cryonics Institute.

Other cryonics organizations exist or are being developed throughout the world, such as CryoRus in Moscow, Russia, and the American Cryonics Society.[13]

To keep this explanation relatively straightforward, we will only discuss costs for Alcor and the Cryonics

---

[13]American Cryonics Society - Human Cryopreservation Services for ...

American Cryonics Society - Human Cryopreservation Services for the 21st Century.

www.americancryonics.org/

Institute. This covers the organizations that represent well over 90% of worldwide cryonics membership.

The reader should be advised that this book is written in 2018 of the Common Era, and that costs generally can (and do) change over time. The purpose of this chapter is to provide an accurate and comprehensive overview of the cost of cryonics as of this moment, and also to deal with future costs in a reasonable, although speculative, manner.

Costs at Alcor
Alcor has two options for cryopreservation—full body and neuro vitrification.

The full body cost is currently $200,000, generally paid with an extra life insurance policy. The cost for neuro vitrification, or "neuros," is $80,000, also generally paid for by life insurance. (These prices do not include the Comprehensive Standby Waiver.)

Alcor recommends you overfund above and beyond current costs in order to handle possible price increases (which will almost certainly occur) as a result of technological enhancements and price inflation.

Early in 2013, Alcor reversed a long-standing policy of "grandfathering in" rates at the costs that were in effect the day that people signed up. There is now an additional dues cost designed to incent people to have CURRENT level funding for when the actual service of cryopreservation is performed.

What does this mean as a practical matter? It means that more coverage is better, quite simply.

But how much more? Well, here are some considerations.

Let's assume, for a moment, that you signed up with Alcor in 1994, the year the author of this book signed up. The cost of a whole-body cryopreservation then was $120,000. If I had to be cryopreserved right now, in 2018, Alcor would provide me with a service legitimately and currently costing $200,000. (This amount of $200,000 charged by Alcor, one should realize, includes about $100,000 which is added to the internal "Alcor Patient Care Trust.")

The new pricing model indicates I will need to have $200,000 designated to Alcor to secure a state of the art cryopreservation. So, if I had only obtained $120,000 of life insurance for my funding when I signed up, I would now need to buy an additional policy to make sure that I was funded for the full current cost of $200,000.

The "Rule of 72" and How to Plan for Your Future Cryopreservation Needs

This short explanation may seem like a digression at first, but we will soon find out it is not.

Have you ever heard of the "Rule of 72?" It is astonishingly simple, but powerful.

You can determine the future cost of a compound interest calculation by dividing the number 72 by the interest rate. The result indicates the number of years it will take to double the original amount. Cool, huh?

For instance, let's say we are buying an item which costs $200,000 today – like a whole-body cryopreservation at Alcor. Let's further assume that medical inflation increases costs by 7.2% a year. While no one knows for certain what inflation rate to expect in the future, you are almost certainly aware of current macroeconomic events which make ongoing inflation virtually inevitable.

It is extremely unlikely, given the actual numbers of people signing up for cryonics throughout the planet, that the economies of scale available for commodities like your iPhone are going to happen for cryonics in the near or midterm future.

Medical Inflation Rates

By way of a data point of hard reality, the health insurance premiums I pay went up 30% and 17% per year for the last two years. So, 7.2%, as a working figure for increasing costs of specialty medical interventions, is not unreasonable.

The number 7.2 goes into 72 ten times. This gives us a rather convenient shortcut for calculating without a calculator or even without pen or paper how costs will rise.

At a 7.2% inflation rate, costs will double every ten years. So, a service which costs $200,000 today will cost $400,000 in ten years. Ten years from that point, the cost doubles again to $800,000, and in just another ten years the cost jumps to a staggering $1.6 million.

However, historically, Alcor's increase in cost has been more modest, coming in at 3.2%. Again, using the "rule of 72" we divide 72 by 3.2, and come up with a doubling time of about 22.5 years.

So, if we are anticipating that the rate of increase in cost for Alcor will continue the historic trend of 3.2%, and we want to make sure we are covered for this future increase in cost for at least the next few decades or so, we will want to have double the current funding requirements, or about $400,000 of life insurance.

Costs at the Cryonics Institute
A full body cryopreservation at the Cryonics Institute is only $35,000, which seems like a dramatic reduction in cost compared to Alcor's cost of $200,000; however, there are some important nuances to consider.

Unlike Alcor's cost, in which the medical stand-by team and transportation logistics are included as part of the package, the Cryonics Institute has "unbundled" the cost of cryopreservation and transportation.

So, what are the current global costs of the Cryonics Institute?

The answer is obtained by adding the Cryonics Institute cost *and* the expense of a separate cryo transport and medical standby organization called "Suspended Animation, Inc." along with the cost of a private air ambulance.

The global cost of the Cryonics Institute, Suspended Animation, Inc., and a private air ambulance is currently $143,000, which is expected to increase. To their credit, the Cryonics Institute portion has been stable for several decades.

Patient Care Trust

One of the distinctions between Alcor and the Cryonics Institute is the amount each organization sets aside for the long-term maintenance and costs involved in taking care of the "patients." (Yes, in the world of cryonics, the members who have been pronounced legally dead are referred to and treated as "patients," not corpses.)

Alcor, charging $200,000 for a full body cryopreservation, places a total of about $100,000 in the "Patient Care Trust" each time a patient is cryopreserved. As of 2018, there is around $12 million in this fund. The Patient Care Trust has its own separate board of directors and is not utilized for day to day operation of the Alcor organization, but instead is used strictly for costs of maintaining the patients.

The Cryonics Institute, charging much less, clearly cannot set this much aside for each patient, and they do not. The amount added to the Patient Care Trust at the Cryonics Institute is currently about $20,000 each time someone is put into cryopreservation.

Dues

In addition to the cost of the cryopreservation, which is generally paid by a life insurance policy at the time of death to the cryonics organization, both Alcor and the Cryonics Institute have membership dues.

Membership dues enable the organizations to keep up with the costs of doing business and to be stable and sustainable.

The annual dues (2018) for Alcor are $525.

The annual dues (2018) for the Cryonics Institute are $120. Additionally, at "CI" one can pay a lifetime membership, a one-time cost of $1,250, eliminating any further annual dues.

There is a reduced dues structure for additional family members for both organizations.

It is important to understand that these "dues" are separate from and additional to the cost of the life insurance. These membership dues are payable only prior to cryopreservation. Dues stop once cryopreservation has occurred, since the Patient Care Trust covers all ongoing costs.

Sample Membership and Life Insurance Costs

How are people paying for the costs of cryonics? Here are some actual client examples.

Most clients are paying with some version of life insurance.

Life insurance comes in different types, and the cost varies dramatically with age and the types of coverage.

The following are actual rates paid by genuine human beings. The names have been changed, of course, but these are accurate and representative exemplars of cryonics funding.

John

John is a software engineer, age 29, in excellent health. He invests $142 a month in an Index Universal Life, because he wants an insurance policy that will be in place in his later years without cost increases. He also knows he wants and needs a systematic, disciplined, long-term savings plan, and the Index Universal Life provides a tax advantaged long-term savings plan.

He has an immediate death benefit of $300,000, which is more than sufficient for Alcor's full body cost for now and for the mid-range future. Additionally, John's Index Universal Life policy grows cash - cash which accumulates tax free and is creditor proof. John's policy projects to a substantial sum in the future: at age 100, the cash value *and* the death benefit of his policy grow to over $1.7 million.

Nancy

Nancy is an emergency room nurse and deals with the fragility of life daily. She sees first hand nearly every day how life can provide unexpected and life-threatening events. Even though she is only 37, she would like the security of knowing she is doing everything she rationally can to live on into the future, so she is interested in cryonics.

After filling out a website form for some quotes and information, Nancy is pleased and surprised that she can, indeed, afford the global costs of cryonics.

Although she has a savings plan at work, Nancy likes the idea of an additional long-term savings plan, a way to automatically "pay herself first" a portion of the money that goes through her hands.

She finds that for only $151 a month, she can purchase an Index Universal Life policy for $175,000. That amount is sufficient for the Cryonics Institute and the medical transport team costs. Additionally, the policy enables the growth of a cash value. Nancy does not need to die to benefit from this cash value in the policy, which is growing at a rate of return that historically beats inflation.

When Nancy reaches the age of 100, this cash value, using historic rates, could grow to well over $1.2 million.

Dan

Dan is a PhD student, living on a stipend provided for his graduate studies of $12,000 a year. He is intrigued by the idea of cryonics, but his initial thought is that he probably cannot afford to sign up now.

However, after listening to a motivational tape encouraging him to take some action in the direction of his goals, Dan decides to do some research on different ways of funding cryopreservation.

He is pleased and surprised to find that there are two basic types of life insurance. We'll quickly summarize these provisions here and go over them in more detail later.

"Permanent" life insurance builds an internal cash value, enabling the policy to remain in place in the later years without cost increases, or even become "paid up" with no further premium required. This is the type recommended by cryonics organizations because it can stay in place in the later years when humans tend to die.

There is also "term" life insurance, which is life insurance for a term or specific period of time. Both the life insurance cost (the premium) and the death benefit stay level for a term of time, for example 20 years.

Dan discovers that term life insurance is quite inexpensive in the early years! He recognizes that it can be very expensive in the later years when the statistical odds of death are much higher.

Dan finds out another important fact about *some* term policies. They can be obtained now and can be changed or upgraded to a permanent policy in the future, without evidence of insurability. (No medical exam would be required to upgrade these policies.)

So, after thinking about his age and how the costs may rise in the future, Dan decides he would like $400,000 of coverage.

And he finds he can get this $400,000 in a 20-year level term format for only $344 a year, which is less than he is spending for his mochas at Starbucks over the course of a year!

As Dan continues to investigate, he learns that both the coverage premium and the face amount of the 20-year term policy are contractually guaranteed to stay level for the full 20 years.

At the end of the 20 years, the premiums will take a very large increase. And, because there is no internal cash value to make payments, if Dan is late or misses a payment, term insurance lacks the "robustness" associated with permanent life insurance.

He also finds this policy is "upgradeable." Dan's first thought is that this means he can obtain an increased amount of life insurance with no evidence of insurability. But, as he researches, he finds the "upgradeable" feature means that he can actually change the *type* of coverage from term to a permanent policy like Universal Life or Whole Life with no evidence of insurability.

Dan learns that this can happen at any time during the first 10 years of his 20-year level policy. The upgrade to a permanent type policy will be priced at the then attained age, but no evidence of insurability will be required.

Dan sees this as the perfect and affordable way to enable him to sign up for cryonics now, lock his good health and insurability in, and not face the risk of being unable to obtain a life insurance policy later due to health concerns that could arise.

Lucy

Lucy is a 45-year-old real estate investor. With an eight-digit net worth, she could actually afford to pay the cost of cryonic suspension with cash.

However, as she investigates, she finds that the cryonics organizations do not allow funding to be structured by simply naming them as beneficiaries in a will or even a trust. Alcor, for instance, requires that the $200,000 be set aside with them, growing in a safe money market account. Money market accounts are traditionally among the lowest yielding investments.

Lucy has seen a lot of ups and downs in her career, and she knows the uncertainties in the business market. She has seen colleagues with multi-million-dollar estates unable to pay their bills after major disruptions in the business world. So, she likes the idea of getting her cryonics paid off in a manner that will not require future payments.

But, she understands that her investments average a reasonable rate of return, and to escrow the $200,000 with Alcor, as required, means that she would not have that money growing for her elsewhere. Lucy understands this as the "opportunity cost" which is the simple fact that choosing to invest money in one place means you are not putting that money in other options.

As Lucy continues to investigate, she is pleased to find that she can indeed gain the certainty of fully paid off cryonics funding. And this can be done while still using the leverage of life insurance.

Lucy can reposition a single lump sum of $65,101 and immediately create a fully "paid up" policy, establishing an immediate $200,000 death benefit to pay for her cryopreservation at Alcor.

She finds she has other options. She could create a "fully paid up" policy by paying for the policy over a seven-year period, paying $12,374 a year for seven years. She could, alternatively, pay a lifetime annual payment of $2,664 each year.

All of the above options build up substantial cash value that can be withdrawn from the policy.

Lucy realizes that she can have the certainty of payment without tying up $200,000 at an unacceptably low growth rate. Like other investors and smart individuals, Lucy expects to average 5 to 10% a year or more on her money. At a 10% rate, we could say the opportunity cost of Lucy tying her $200,000 at Alcor is $20,000 a year. How silly would it be for Lucy to pay a $20,000 opportunity cost per year if she can create the necessary funding for under $3000 a year?

Of course, this assumes that Lucy is insurable and able to qualify for life insurance. If she were to be uninsurable, she could still utilize the benefits of a guaranteed annuity for funding her cryonics. This annuity option does not require evidence of insurability, but it does require that the entire amount of the funding cost would need to be repositioned into the annuity.

Wes

Wes is a retired software engineer. At a youthful age 70, Wes does everything he can to stay healthy. However, he realizes that no amount of exercise, diet, or good lifestyle choices have enabled his friends to beat the inexorable encroachment of aging and involuntary death.

He notes that professional athletes, astronauts, vegetarians, marathon and ultramarathon runners, anti-aging activists, and even self-styled health gurus are still dying at rates not far from the mortality expectations of the general population.

As an engineer, Wes believes that the diseases of aging are best described as a very complicated engineering problem. He has been following the advances and breakthroughs in anti-aging medicine closely. He is excited about the emerging science of regenerative medicine, with the promise that stem cells, cryopreserved organs, organ printing and banking, and other interventions will enable dramatically extended health spans for intelligent humans who can take advantage of these technologies.

Wes realizes that he is quite possibly in the last generation in which involuntary death and the horrors of aging are inevitable.

Wes has good reason to believe that future technologies will enable intelligent humans to live well past the current life expectancies. He continually discusses this with his friends, most of whom dismiss his obsessions with aging and defeating death as wishful thinking.

But Wes reads widely and cultivates an ability that served him well as an engineer. He continually keeps the big picture in mind, while being aware of subtle details. This ability to visualize the arc of history has given Wes a different view of life than his more provincial friends. While his contemporaries look back fondly to the so-called "good old days," Wes has a completely different orientation.

He reads books like *The Rational Optimist: How Prosperity Evolves* by Matt Ridley, and *Abundance: Why the Future is Better Than You Think* by Peter Diamandis. These books document how, by virtually any metric and for the vast majority of humans, life is better than it has ever been. And humanity is rapidly learning how to create a thriving civilization that will enable humans and all sentient beings to live not with pain but experience a continuum of bliss.

Wes does not want to miss this!

Despite the good-natured ribbing he takes from his friends and contemporaries, Wes is pretty sure he is not delusional. By applying the standards of health and longevity that were in place as recently as 1900, he and virtually everyone he knows would already be dead. Zooming back to take the long view of history, Wes can see the long-term trends while his grumpy friends focus on the short-term and negative perturbations.

Is it too late for him?

He needs a bridge to get him across to a future technology. After all, it's amazing how many advancements have taken place in his lifetime, particularly the last decade. He feels that it's not a question of *if* this technology will exist, but rather *when* it's going to finally happen.

Wes had heard of cryonics, but assumed, as many people do, that this is a fantasy technology, and even if the technology were available, it is only for the mega-rich and powerful.

And, at age 70, surely this is financially out of the question for him. While healthy, Wes also has some cholesterol issues and takes two different blood pressure medicines.

However, being a possibility thinker, Wes decides to investigate the cryonics option.

He finds he can fund membership at the Cryonics Institute, which would also include private air ambulance transportation and the specialized medical transport team, Suspended Animation, Inc. And the global cost of these is just $143,000. Providing a cushion against cost increases, he figures he needs a minimum of $160,000 of life insurance coverage.

He finds he can obtain this $160,000 of permanent coverage for an annual payment of $7,734.

Wes is on his second marriage, and his current wife, Susan, is supportive of most of his plans.

However, since they pool most of their funds for living, Susan is understandably concerned that her husband might spend money on unproven or speculative ventures.

While they had discussed the idea of cryonics in general in the past, Susan had not expected that Wes would pursue the idea seriously. Now here he is, sitting across the kitchen table, asking her to figure out how to put an almost $8,000 a year budget item into their plans!

"It is a matter of value systems!" exclaims Wes, a little more stridently than he had intended. "We have friends who put much more than that amount into boats, airplanes, fancy cars, country club memberships, and optional vacations."

"Yes, Wes, that may be true, but boats, airplanes, fancy cars, and vacations are all things I can brag about to my friends. They have value that can be seen by anyone who is paying attention. Instead, you want to take this money and put it into something no one can see. Could this be just a little selfish and not quite fair to me?" she asks.

"Well, let's think about this, Susan," observes Wes, realizing that he is entering one of the most crucial conversations of his life. "What if I had a heart condition or other life-threatening disease, and the cost of keeping me alive was less than $8,000 a year? We have both worked hard and become reasonably successful. If I needed this money to live, or if you required this $700 a month for your required medications, we would find it. How would you feel about selling the RV that we have not been using much? That would free up about $600 a month. Now we are just talking about a hundred a month, and I can save that with one or two less rounds of golf."

"Wes, I'll tell you what; if you handle the details of selling the RV, and we look at our budget and see where you, not I, can redirect some spending, we may be able to make this work. But, Wes, I want you to remember that cryonics is your dream and goal, not mine. I will support you and do what I can to get you the best cryopreservation possible, but don't expect me to sign up, even if we could afford to have both of us signed up."

"Oh," she continues with a sly smile, "and the next time I want to do something, and you don't, we aren't even going to have a discussion. We'll do this for you, and the next big decision, you will happily go along with my wishes. Whatcha think?"

Wes is delighted with his wife, the mature relationship that they had developed, and that he could, at long last, pursue a technology he had read and thought about for literally decades. He had read numerous science fiction stories that used the basic premise of cryonics, but to find out that cryonics was a real option and now within his grasp financially, is exciting. While he understands that cryonics provides no guarantee of success, as a practicing rationalist, he believes that this is the most intelligent course of action he can take to have a chance to see the year 3000.

The above real-world examples show how actual human beings, very similar to you, in a variety of financial situations, have been able to fund their cryonics arrangements.

Chapter 4
Jerry's Epic Cryonics Odyssey

In this chapter, we will meet Jerry, another potential time machine traveler. He is a fictional character, but as a composite of many real cryonics clients, Jerry's story will help us further understand issues involved in cryonics and its funding.

Jerry Reynolds is a 45-year-old software developer. (About 60% of current cryonics signups are software engineers or in related fields.) Like many people, Jerry first heard about the cryonic suspension option from a co-worker friend.

Jerry went home after hearing about the concept and put "cryonics" into his internet search engine.

He was surprised by the number of responses that were elicited by such a simple search. His friend had mentioned both "Cryonics Institute" and "Alcor," and Jerry clicked on Alcor. Although he used software search engines every day, he still marveled at how easy research has become using the internet.

Jerry found the Cryonics Institute website and the Alcor website were both comprehensive and professionally executed. As he examined the Alcor site, there were pictures of the facility in Scottsdale, Arizona, along with pictures of some of the people who were involved with the organization.

Jerry was especially interested in the "Frequently Asked Questions" section, where some of his exact questions and concerns were dealt with.

"This is fantastic," thought Jerry to himself. "This is more than a concept; there are several actual up and running operations!"

As Jerry read on, he found the general information on costs particularly interesting. Alcor was charging $200,000 for full body suspension. And there was a "neuro vitrification" option, with just the head frozen, that was $80,000. Additionally, Alcor was recommending a "cushion" of at least $50,000 and ideally much more for potential future upgrades and possible extra "standby and transport" costs.

Immediately Jerry's mind went back, unbidden, to the constant financial struggle that always seemed to be on his mind. "I can't understand why we feel so pressured financially all the time," he thought to himself. "I have some advanced education, and I am willing to work hard. We keep getting raises, but the money just never seems to be enough. Even with my wife, Pat, working, we don't have the financial resources for something that might cost

$100,000 or $250,000. I am working my tail off, and I made $86,000 last year, with Pat bringing in another $50,000 from her hospital administrator position. This cryonics thing must just be for really rich people."

Now here he was considering a procedure that would cost a sizable amount of money. Yet he was thinking, "This is something I really might want to do. What should I do?"

Jerry could just imagine the response his wife, Pat, would have to the cost issue of the cryonics thing. He had met Pat in college in 1982. They had been married for 17 years and had a great family. He dearly loved his wife and respected her keen wit, broad intelligence, and sparkling personality. And they still found each other romantically and physically attractive. *But* he also knew his wife inside and out! Presenting a cryonics option to her was going to be a challenge.

He chuckled to himself as he thought about how ridiculous it would sound to Pat if he said, "Listen, Pat, I'm seriously looking into taking our retirement money and the kid's college funds. I am going to liquidate these during one of the worst stock market dips since the 1970's. And I am going to use this money to freeze myself when I die."

No, this definitely would not do! Jerry was disheartened and discouraged as he thought about how to create the funding for what he was already regarding as "his suspension." It seemed funny. A few days ago, he did not even know that such a possibility existed. And now he found that it *did* exist but was not financially feasible. He was angry. "It's not fair," Jerry thought. "This thing should not cost so much! What kind of a rip-off is this, where these cryonics people only freeze elitist, egghead, rich people?"

Jerry went back to the Alcor website. As he continued to read, his initial frustration and anger turned to hope.

It turns out that most cryonicists fund their suspension with life insurance!

Jerry knew that life insurance created a big lump sum of money when you died, because his father had died of a heart attack at age 57, when Jerry was only 27. Jerry even recalled that the policy proceeds were not subject to income tax. He also remembered how part of the fear and apprehension his mother had about the future left her eyes when the insurance man delivered her a check for $175,000. That was considered a very large policy at the time, as Jerry recalled, and it had certainly made a difference to their family.

As Jerry continued to check out the Alcor website, he noticed that there was a page with links to insurance agents familiar with the cryonics business.

"Now we are talking!" thought Jerry as he clicked on the first of the six hypertext agent references that showed up blue in his browser. As the box opened for him to write a letter, he wondered what he should say.

Finally deciding that a direct and simple approach would be best, he wrote the same letter to several of the agents listed in the Alcor website.

"Dear Sir or Madame,

I am looking into funding a suspension with Alcor. I would like a quote for $200,000 of coverage."

Then he put his date of birth and added that he was a healthy nonsmoker. As an afterthought, he also added his height and weight. Jerry was proud that he had kept reasonably fit and wanted to get the best possible rates.

Feeling like he had really accomplished something meaningful and was actually moving forward and gathering information on whether he could afford cryonics, Jerry was rather proud of himself. Feeling pleased that the ball was at least rolling, Jerry turned the computer off and went to the kitchen to spend some time with Pat and the boys.

The next day when Jerry got a break at work, he anxiously checked his email. Sure enough, one of the agents had written back. Jerry opened the letter quickly, since he was getting quite enthusiastic about the possibility of a cryonics option.

Taking a quick peek around the office to make sure none of his nosey co-workers were monitoring him, Jerry read the email.

"Dear Jerry,

Thank you for your interest in funding your cryonic suspension with life insurance.

Based on the information you provided, there are several ways we can help you. As a 45-year-old man in excellent health and height and weight ranges, you will probably qualify for the "preferred" rates.

To provide the recommended "cushion" and fund for a whole body or "best available" suspension with Alcor, we will need to create a lump sum of at least $250,000.

The following quotes will be with an A-rated carrier that has been around since the 1800's. More importantly, this insurance carrier has put in writing that they have no problem with cryonics organizations as beneficiaries and owners of the insurance policy.

We can do this with a 20-year level term life insurance policy for $432 a year. This rate will stay constant for 20 years, with the premium (cost) of the coverage staying level at this rate, and the death benefit (face amount of the coverage) also staying level.

This term insurance also provides guaranteed upgradeability to Universal Life or Limited Pay Life.

If you prefer, you may want to own a permanent policy like Universal Life, in which the premiums stay the same past age 120, and the policy accumulates cash value.

If you invest $223 per month into an Index Universal Life policy, you not only will have the $250,000 of "death" benefit, but a cash accumulation while you live as well. We can go over other quote amounts too, when we talk.

The cash value grows inside of the policy - tax deferred and free from the predations of creditors.

If you get sued, if you file bankruptcy, if you have creditors of any sort, or may in the future, you should know this: Any money you have in stocks, bonds, bank savings deposits, mutual funds, money market certificates, or certificates of deposit at the bank; ALL of these are subject to the claims of creditors. But the cash value in your life insurance policy is NOT. In most states, Jerry, cash values of life insurance are NOT available to creditors, even in bankruptcy.

The cash value grows at a good rate of return, currently about 7%, which is higher than the rate the banks are paying for savings and even CD's. And the money is safe and guaranteed by the insurance company.

Jerry, I would like you to email me, or we could have a real-time phone/video call. Let's talk about what will possibly be the best match for you, given your individual circumstances. Once we have a chance to correspond and we find out what might work for you, I will get you a computer illustration out by email or overnight mail.

Warmly and professionally yours,

Rudi Hoffman

"Well," thought Jerry, "This guy has given me a lot to think about. He obviously understands what I am trying to do and sure seems to know what he is doing."

"But I've never really understood life insurance. And I don't really *want* to understand that much. I just want to know I am not getting screwed by buying a bad policy. And, I need to be able to explain to Pat that I have done enough homework, so I can help inform our decision making."

Jerry continued his internal dialogue, "I know we really like Stan, and he's done a good job on our house and car insurance. And we are in Kiwanis together. But this cryonics life insurance issue is a specialty. I am almost certain Stan wouldn't get this cryonics thing. And I see here where I can book an appointment online to talk directly with this Rudi Hoffman, so I think I'll do that."

"And the good news?" Jerry reflected happily. "I can do this thing! Even the most expensive policy this guy emailed me was less than three hundred bucks per month."

"That sure is a lot better than trying to come up with a lump sum of $250,000! And, it won't take away from the money that I want to leave for Pat and the boys."

Jerry made an electronic file for this information, opening it under "cryonics life insurance," and resolved that he would think more about this tonight.

The next day, Jerry emailed rudi@rudihoffman.com and utilized the online calendar function to book an appointment for a phone conference with Rudi Hoffman.

Then he excitedly began researching life insurance and his options. To his surprise, he found some fascinating information, some of which is coming right up in Chapter 5!

Chapter 5
A Non-Boring Chapter on Life Insurance

The first recorded life insurance policy was issued in London on June 18, 1536, on the life of William Gybbons. It was a one-year policy.

And Gybbons died within that year! On May 29, 1537, Gybbons died, although there is no written cause of death. Perhaps you will not be surprised that even back then, there were unscrupulous companies. The insurance company refused to pay the death benefit! The position the insurance company took was this: by their reckoning, Mr. Gybbons had life insurance for a year. They insisted that Gybbons had lived a full year by their calculations, using a year as defined by twelve months of four weeks each.

Fortunately, the court ruled against the insurance company, and they were forced to pay.

Perhaps with such an inauspicious (or should we say suspicious?) beginning, the institution of life insurance is viewed with wariness by many (okay, perhaps most!) people.

But the insurance industry continued to grow and eventually became extremely regulated.

And, due to the competitive pressures of both the free market and greater regulation, the policies that are available today are exponentially better for the consumer than policies issued in the past.

As an example of how things have improved for the consumer, the life insurance policies issued after 1995, which accumulate so called "cash value" or "living value," have cash accounts growing at dramatically better rates of return than older policies.

Types of Life Insurance Policies
Basically, there are two types of life insurance policies.

Broadly, these are "term" and "permanent" policies.

Included in the "permanent" policy category are policies which are designed to have level premiums in the later years. These policies have a cash value which enables the premiums to stay at the original cost or even to stop altogether. Permanent policies go by names like Whole Life, Universal Life, and Index Universal Life.

Term Insurance

But let's start with the first and easiest to explain coverage, "term" life insurance. As the name would suggest, in term life insurance, both premiums and coverage are level for a "term" or period of time. Pretty straightforward so far, don't you think?

Let's say you are a banker, and a real estate developer comes to you and borrows $10,000,000. As a banker, you have reason to believe that the developer will be able to pay debt service on the loan. But, if Mr. Developer dies, his project could fold, and the loan could default.

It is prudent (and most bankers will require) that a life insurance policy be taken out on the life of Mr. Developer. Because the period of liability is known in advance and the loan will be paid off over a specific period of years (for example 30 years), and the face amount is very large, a term policy may be appropriate in this situation.

It is not that the insurance company thinks that the risk of you dying is actually going to be level for the next 30 years! According to insurance company statistics, your actual odds of dying are about ten times as high at 55 as they are at age 25.

So, what the insurance company does is "level" out the premiums, so that the premium cost and the insurance stay level for the term period.

Because the term policies are for a limited period of time, the insurance company has accepted the risk of paying a death benefit should death occur during the period of coverage. So, a term policy may provide coverage for a 20-year term, but if the policy is not *renewable* and death does not occur during that 20-year period, then the company has no risk of paying a death claim at the older ages when death is more likely to occur.

The good news is that most term insurance today is guaranteed renewable. This means that the policy can be renewed as term insurance at the end of the term. However, the rate, or premium, will go up substantially. In some cases, the premium will increase 15 or 20 times, or 1,500% to 2,000%. This increase in the later years is why life industry statistics document that only about 3% of term policies result in the face amount of the policy being paid out in a death claim.

Many companies offer term insurance that is also *upgradeable*. This means that the policy can be "upgraded" to a Universal Life or a Limited Pay Whole Life policy with no evidence of insurability (no health checkup required).

One major concern with term life insurance is that, upon the end of the initial level period, the rate will go up. And not just go up a little bit; it could increase 20, 30 or 50 times the initial rate. This explains why 97% of term policies do *not* result in a death claim. The increasing cost in the later years, when people tend to die, disincents people from renewing the policies. And the result of this nonrenewal of the term policy means the life insurance company is off the hook altogether!

Permanent Policies

So far, we have covered term policies.

You are doing great. I'm sure everything has been crystal clear so far, and you might even think that you won't really have to put your thinking cap on to learn all you need to know about life insurance.

Well, I apologize in advance for this, but I must tell you modern policies have morphed, evolved and improved. In doing this, however, they have become more complicated, and you almost certainly will need to put that thinking cap on to understand the state of the art status of permanent policies.

Please hang tough with me as we magically fly through a basic tutorial on permanent policies. These policies are designed and solidly engineered to actually be in place on the very day you absolutely need your policy to be in place. That would be the day that your faithful heart stops beating. (And, hopefully, there are cryonics technicians by your side to send you on your ambulance ride to the future.)

In general, permanent policies stay in place until you die.

This is accomplished by charging you a higher premium than the term insurance we learned about earlier. The extra premium allows the insurance company to build a cash value in the policy and keep the premium cost level your whole life - even though statistically, you are more likely to die in your later years.

The cash value also enables a *living* benefit to the policy. In other words, *a part you don't have to die to access.* This cash value can be accessed one of two ways: 1. You can surrender the policy and give up the coverage, or 2. You can borrow against the cash value, keeping the policy in place.

Traditional Whole Life: Many people (yes, including the author's sister) still use the term "Whole Life" insurance synonymously with "Permanent Life" insurance to describe any policy designed to stay in place until death. This is no longer technically accurate, since "Permanent Life" now includes a much wider range of insurance policy types.

Traditional Whole Life insurance was designed to stay in place your entire life without a premium increase. It was the first iteration of life insurance that didn't have the problem of the policy dying *before the client did!* The problems it DID have, however, included being very expensive and paying low returns.

Traditional whole life was vilified by consumer advocates as a terrible investment that was ridiculously expensive. The life insurance industry has always been competitive and rose to the challenge. It responded by evolving more affordable, more interest-sensitive products with greater flexibility for clients.

The newer forms of policies pay a higher growth on the cash value part of the policies. Some policies, called *variable life policies,* enable the cash value to be directly invested in mutual funds. Others, like Universal LIfe and Index Universal Life, are tied to the market and allow consumers greater chances for substantial growth.

Universal Life: Universal Life is an innovation that functions in some ways like "buy term and invest the difference," but in a single agreement that means the "invest the difference" part in the cost between term and Whole Life actually does get saved. And the structure of the policy enables a lower cost of insurance component than term purchased separately.

In a Universal Life policy, the buyer has more flexibility of payment than in traditional Whole Life policies, and the cash value grows at rates determined by a prevailing fixed rate indicator. The cash value growth rates generally track a bit higher than those available in bank certificates of deposit (CD's) rates.

Index Universal Life: Relating to cryonics, Index Universal Life has combined the most beneficial components of all its predecessors. *This is the king of the jungle!* Here's why: In Index Universal Life, the interest rate credited to the policyholder can be tied to the growth of a stock market index. Unlike with variable life, however, a loss in the index does not reduce the cash value. Instead, the insurance company provides a guaranteed *floor* of 0 or maybe 1 percent interest, even if the stock market index is negative. Talk about a win-win!

How can an insurance company do this and still provide a sustainable program? There are always tradeoffs, and the tradeoff in the Index Universal Life is that there is a cap (or maximum) to the index amount which is credited to you. For instance, with a 14% cap you get 100% of what the Standard and Poor 500 index does in the year up to a maximum of 14%.

Quick bottom line? This kind of policy is really good for you as a consumer (and potential cryonicist).

This design enables a very good growth rate to be credited, while providing greater safety of the cash accumulation in the policy.

I promised you I wouldn't be boring. The following two paragraphs are only for those who want more details on this amazing innovation called Index Universal Life.

The Index Universal Life illustrations (which show what could happen in your policy) also have at least one section showing a "quadruple worst case scenario," in which the market returns are continually negative in addition to a second rather unlikely scenario. This is briefly summarized as a situation in which everyone is dying of a worldwide pandemic, which is pretty much the only circumstance that would require the insurance company to go to a "fallback" rate of higher internal charges and internal costs of insurance.

The reasons behind why ULs and IULs have these ultra-worst case higher internal cost sections has to do with required reserves and government regulations and goes beyond our attention span here, I promise. But, suffice it to say, the practical outcome of these variables is that we have this genuinely great improvement in what a policy can do for us as life insurance/investment consumers. The challenge is that the illustrations showing what the policy will do for us are 18 to 50 pages long! Holy moly! Add to this the application paperwork of another 12 to 30 pages, and one begins to understand why even very smart and analytical people become resistant to detailed policy analysis.

All readers welcome back here!
The summary and take-home story here is that new policies, like Index Universal Life, are advantageous for long term affordability. But don't be surprised or intimidated by the many illustrations included with your proposed contract. Since the insurance companies can only project future returns, they are legally and ethically bound to give you information on different potential future financial scenarios, including possible worst cases.

A quick analogy: a car and a bicycle are both transportation tools, but a car has more moving parts and options. If you are traveling cross-country on a life or death mission, you're probably going to want the car over the bicycle, even if it does have more parts and a different price point.

Common to Both Term and Permanent Policies

There are some features of life insurance which generally apply to both term and permanent types of policies. Two of the most important provisions relate to contestability and suicide coverage.

Contestability Period

Nearly every life insurance policy has a clause called the Contestability Period. Here is exactly what the verbiage of the policy sitting on my desk states:

"Contestable Period. During, but not after, the contestable period, we can contest the validity of the new policy and reduce a claim for any misrepresentation or nondisclosure of a material fact in the application for exchange. The reduced amount will be that which the premiums paid on the new policy would have purchased on this policy had the exchange not occurred. The contestable period starts when the new policy goes into force and ends when the new policy has been in force during the insured's lifetime for two years from its policy date."

"Whew!"

(Doesn't my writing read better compared to that?)

Exactly what does this mean? It pretty much means this: If you lie on the application and die in the first two years the policy is in force, the company may pay a reduced amount or not pay at all on your claim.

Is this a big deal? Not really. First of all, the rather obvious solution is not to lie on your application. And, as it turns out, it is pretty difficult to make a major and significant misrepresentation that is not discovered during the underwriting process.

For instance, the most common misstatement on a life insurance application, according to most underwriters surveyed in an informal poll, is "fibbing" about tobacco or nicotine usage. But, the life insurance company does not merely take your word on the life insurance application that you are not a user of nicotine products. They have a nurse come to your home or place of business and take a small blood and urine sample.

The tests run on these samples are very sensitive and will determine if any nicotine usage has occurred within a period of several weeks. These tests also serve as screens for AIDS and HIV virus, lipid (fats) and cholesterol levels, and protein and sugars in the urine. Even if you have seen a doctor or have had a checkup recently, most insurance companies will ask you to have the urine and blood tests on-site with the nurse that they hire. This way, they have more positive control of the "chain of custody" of the samples and are less likely to make an error in determining who they will cover and at what rate.

So, we see that a "material misstatement" on an application is unlikely to cause a claim to be contested. While an intentional and fraudulent application might have the possibility of sneaking under the radar of the underwriters, causing a claim to be contested should death occur in the first two years, this "fine print" should not concern any of you good, honest readers.

The Suicide Clause

Do you ever watch mysteries on television or at the movies? Or perhaps you are a fan of murder mystery novels, and like to figure out "who done it?"

In many of these plots, people go to great lengths to determine if the cause of death was suicide, and one of the plot points is that policies do not pay in the event of suicide.

Well, guess what? We can add this to the list of popular cultural assumptions that are incorrect! In most cases, life insurance *will* pay, even in the event of suicide.

To be fair and accurate, this was not *always* the case, and some older policies, written before 1970, still have provisions against paying the death benefit in the event of suicide.

Nearly all large and reputable companies now have what is called the "Suicide and Contestable Period." This is a period of time, generally two years, in which the company can contest or legally not pay a death benefit if they determine you made major misstatements or if you commit suicide.

But, after the policy has been in force for two years, the insurance company will pay even if you kill yourself! How remarkable is that?

A Heartbreaking Personal Story: How the Suicide Clause Worked in Real Life

As a personal aside, one of my first death claims, several decades ago, remains forever etched in my mind. The phone rang, and it was one of my clients. He had referred me some time earlier to his daughter, who had purchased a fairly large life insurance policy on her own life. His daughter, 28, was an attractive waitress in vibrantly good health who had qualified at the preferred rating category.

Here is how the conversation went.

"Rudi. I can hardly say these words. My daughter is dead."

"Oh, my goodness, Mr. Smith! I am SO sorry. What happened?"

"She killed herself. She left a short note, apologizing to us and to her 3-year-old daughter. Then she went into the garage and turned on the car with the garage door shut. We found her yesterday morning."

"Rudi, will that life insurance policy you sold her pay the claim? You will recall that her mother and I are the beneficiaries, because her ex-husband was such a scoundrel. But now, besides losing our beautiful daughter, we suddenly find we are going to be raising our granddaughter; and we are desperate to figure out what resources we can put together to make it work for that wonderful little girl."

When you are a life insurance broker and you get a call like this, you have to be very careful about what you say to the bereaved family. You don't want to say the company *won't* pay if it turns out they will pay. Conversely, you don't want to say the company *will* pay the claim if later investigation and research determine the claim will not be paid. There was a long and pregnant pause in the phone conversation.

"Mr. Smith, let me make a call and find out the status of her policy, so I can answer your question properly. I will call you back in about fifteen minutes."

I immediately called the insurance company. It turned out the policy was 28 months old. This was four months past the suicide and contestability period. And, yes, the policy was in force, and yes, the death claim would be paid.

So, because I was indeed able to deliver this claim, the parents had the financial resources to provide for their granddaughter. While this is far from a "happily ever after" story, it does have the benefit of being absolutely true and deeply meaningful to me personally.

The story dramatizes the remarkable fact that life insurance policies from reputable companies do indeed pay claims, even for suicides, after the policy has been in force for two years.

"Yes, but how might that affect me?" you might ask. "I have no plans to kill myself."

One reason this two-year suicide and contestable period might be especially important if using life insurance as the funding vehicle for cryonics is this: what if you get a terminal and debilitating disease that will corrupt your brain and mind?

It is not out of the question to consider that you may want to maintain the option of some sort of intentional or assisted "death." While the current legal and regulatory climate makes this action difficult to actually carry out, the attitudes and the laws surrounding suicide or assisted suicide may change in the future.

It should also be noted here that cryonics organizations have *no* desire to become embroiled in a potential controversy associated with suicide. Consequently, the cryonics organizations have rather strict protocols and internal controls designed to minimize legal and ethical problems that could arise in the event of an intentional self-inflicted death.

Meanwhile, it is good to know that your insurance policy has suicide and contestability clauses that specify they do not contest claims for those reasons after the policy has been in force for two years.

Back to Our Story

If you have read this far, you have already learned a lot about the exciting world of life insurance. And now you deserve a break. Let's rejoin our hero, Jerry, as he continues his odyssey of researching the cryonics option.

You will recall he has filled out a website form and will be getting some quotes. We join him as he is getting ready for the Christmas holiday.

Chapter 6
Is Your Life Better than Grandma's?
(Why You'd Want to Be Revived in the Future!)

The next day was reserved for decorating for Christmas. While Jerry did not believe every story and fairy tale about Christmas, he did enjoy the celebration, the lights, the decorating, and the cultural phenomena that made up the whole holiday season.

His favorite grandmother, his Dad's mother, Jane, was here helping put ornaments on the tree. Jerry always found this activity nostalgic and meaningful, as he recalled earlier Christmas seasons. Grandma Jane was getting up in years, and clearly slowing down a bit at 91, but her mind was still remarkable. She had been a college professor most of her life, with a background in public advocacy for quality public education. She also had an astonishing memory and would often quote large passages from the books she read. Jerry remembered her legendary library, and the many happy hours he had spent in that quiet corner of the house curled up with one of her books. Jerry loved Grandma Jane for many reasons, but especially for her wide-ranging erudition and astonishing breadth of knowledge.

A Visit with Grandma

"Tell me about your earliest memories, Grandma Jane," requested Jerry. "Let's let this Christmas tree sit for a while, admire our work, and have a cup of tea. I would love to hear you talk about how life has changed during your lifetime."

Grandma Jane sat with surprising grace, taking the offered cup of tea from Jerry. It was just Jerry and Grandma Jane in the room now. The rest of the family was off shopping, and Jerry had wanted to have this conversation with Grandma Jane for a long time.

"Grandma Jane, you were born in the early part of the 1900's. Can you tell me about the changes you have seen? Do you think that humanity has made progress? Is life getting better or worse for most people?"

"Well, Jerry, I am glad you asked these questions, because my answers are probably different than most of my contemporaries. It seems that everyone I talk with and all the television news and media seem to talk about is how violent and uncivilized we have become. Most people seem to think that our life has gotten worse. Grandson, I am here to tell you, as someone who has been around a long time, that the reality is exactly the opposite."

This was surprising to Jerry. He had expected the usual homage to the glories of yesteryear which his other grandparents always provided. "Do you really think so, Grandma Jane? I mean, look at the news. It seems every day the news gets worse. We have violence, wars, and mass shootings. Do you think the human condition is better now than in years past?"

Grandma Jane was reaching in her oversized purse where she always kept books she was reading.

"Jerry, I am going to tell you a secret; in fact, one of the most important secrets of my life. But first, I want to explain and provide some solid reasons and documentation about why I am so positive about the future. I have a couple of books here that speak directly to the question that you and I have had many discussions about; basically, whether life is getting better or worse for most people?"

"What is the secret, Grandma Jane?" Jerry was fascinated that his favorite relative had a big secret, and he couldn't wait to find out what it was.

"I promise to tell you, Jerry. But first I want to tell you why I am so convinced that, for most people on the planet, life is, by every metric, simply better than it has ever been."

The Future Looks Great
Grandma Jane pulled some newer, but clearly well-read hardback books from her purse. "Have you ever heard of Peter Diamandis?" she asked.

"Hmm, let me think," responded Jerry. "Yes, I think he's the guy who was involved in those X Prizes, where they give prize money out to groups who are the first to do technologically difficult tasks. Didn't they have a competition a few years ago to see what groups could get a spaceship into near-earth orbit and then relaunch it?"

"That was indeed an X Prize event, Jerry," observed Grandma Jane, making Jerry feel like a happy schoolboy who was the only student who could respond to a teacher's quiz. "And there are a lot more X Prize contests for all sorts of science fiction technologies, like a real-world medical device equivalent to Star Trek's 'Tricorder' and software that enables inexpensive computer tablets to provide education to children in remote villages. But Diamandis is also aware of what he calls the "Grand Challenges" facing humanity. His book, *Abundance: The Future is Better Than You Think*, documents some of the challenges we face as a human race and what is being done to solve them." Grandma Jane handed the book to Jerry.

"Like overpopulation and world hunger? And maybe water shortages and environmental pollution?" Jerry asked as he took the book, reading some of the endorsements on the back cover, wondering where Grandma Jane was going with this and what it could have to do with her secret.

"Exactly, Jerry. It is no secret that we have poverty, overcrowding, shortages of food and water, and violence in various places around the globe. Most people feel, because of the way popular news and media disseminate information, that these problems are worse than they have ever been."

"The surprising reality, Jerry, is that the mass starvations and horrific dystopias predicted by authors like Thomas Malthus and Paul Ehrlich simply have not happened. Turns out that most of the smart guys who predicted the future to be terrible were simply terribly wrong. Not only are most of us not starving, but of the seven billion or so of us on the planet, about 6 billion are wealthier than humans have ever been. And I am not just talking obvious things like cars, houses, clothes, access to clean water, electricity, good lighting, and savings plans. By more important metrics like life expectancy, health care options, and even a hugely important thing called justice, the average human today is far better off than the average human of fifty or a hundred years ago."

"Jerry, I have been around a few years. Even in my lifetime, I remember going outside to the backyard to get water. Of course, we didn't have indoor toilets with running water, we all had outhouses. Going into town, only about ten miles away, was a big deal, since the cars were unreliable and not a single road was paved where we lived. We were so excited when we got a phone! It was a party line, of course, but it made life so much better. People who carry their cell phones in their pockets these days have no idea of the amount of time and energy they save by being able to communicate to just about anyone on the planet without leaving their chair."

Jerry pulled out his iPhone and observed, "Yes, this device is really a piece of 23rd century technology that most of us can afford. I have heard there are some 7 billion smartphones on the planet, each of them providing access to more information than governments or royalty could have possessed just a few decades ago."

Grandma Jane's eyes sparkled behind her bifocals as she continued Jerry's point. "It is not just that phones have gotten better. We have access to things like computers, flat screen televisions, LED lighting, and transportation options that kings and queens could not have had a few generations ago. Your nice but modest home here is more comfortable than a drafty castle a king might have lived in during the Middle Ages. Even a hundred years ago, most people lived hardscrabble lives of uncertainty and poverty. Lives aren't just longer, most of us are healthier."

"And we are just getting started!" she exclaimed with an enthusiasm that belied her years. "What we are talking about here, Jerry, is the big paradigm shift that most people don't seem to understand, although it is pretty obvious to me. Don't listen to the windbags who keep telling you that the good old days were better! I was there, Jerry. I promise you, the good old days, as the kids say today, really sucked!"

Jerry smiled at his wise and beloved grandma's use of the vernacular. He realized he was thinking of Jane as an intellectual colleague rather than just his favorite relative.

Jane was on a roll. "But here is the other part of the big picture. Not only is life better, richer, fairer, healthier, and more intellectually stimulating, it has more options than it did 50 years ago. *And* the trend is accelerating at breakneck speed. Check out some of the trends that are coming. Diamandis, in his book, documents how progress in artificial intelligence, robotics, infinite computing, ubiquitous broadband networks, nanomaterials, and synthetic biology are exponentially growing technologies that could enable greater progress in the next few decades than has occurred in the past two centuries!"

Jerry was unsure how to respond to this level of enthusiasm about the future from someone as old as his grandmother. And, he did note with amusement that she was sneaking a peek at the book jacket as she spoke to him.

"There are all sorts of emerging technologies outlined in this book, Jerry," enthused Jane. "Things like 'vertical farming' and a technology that turns polluted water, salt water, or even raw sewage into high quality drinking water for less than a penny a liter. And, wealthy tech entrepreneurs are committing their billions and their talent to making the world and the future better."

Jerry was delighted to see Jane's obvious excitement about future lifestyles. It was such a refreshing difference from the pessimism and negativity that seemed to be the natural habitat for most everyone he talked with. Was his grandma crazy? Was this over-enthusiasm about progress just her particular version of senile dementia? She certainly was articulate, and she seemed to have her worldview grounded in documentable facts.

"You know, Jane," Jerry said after a short pause, intentionally leaving out the honorific of grandma, "I think I get what you are talking about, and I know my life probably doesn't seem as hard as yours. You had poverty and privations I can only imagine, growing up in a rural farm. You managed to educate yourself and you pulled yourself up to a career in academia. No wonder you feel life is getting better, because it got better for you. But that is not everyone's experience. While my problems at work and at home may seem small compared to what you experienced growing up, or what maybe a billion humans alive today face, wondering how they can meet even basic needs, my day to day life just does not feel that much better. I am almost embarrassed to admit it, but much of the time, my life is just not much fun."

There was a long pause while Jane absorbed and considered Jerry's words. As a function of her training, or maybe it was patience born from years of working at being genuinely wise, she seemed to hear not just the words Jerry said, but the intention behind the words. She knew Jerry to be a reasonably happy, successful individual. From the outside, most people would think Jerry did not have a problem in the world. But Jane was also privy to the frustrations that Jerry often felt; the powerlessness, the despair, and the depression that even well-adjusted people often feel. Jerry had, over the years, shared all these feelings with Jane through deep and meaningful dialogues.

Jane slowly took a sip of tea, allowing the silence to grow. Jerry was trying to express something deep in the human condition. "Since most of us are far from the poverty and deprivation that comprises most of human history, why aren't we happier? We are living in the Golden Age that people dreamed about for centuries, and many of us feel miserable a majority of the time."

"Jerry, I think I hear what you are saying," Jane began slowly, looking Jerry in the eye with the direct gaze that was part of her uniqueness. "And, I think it is a central question that we need to figure out as individuals. Why aren't all the improvements in lifestyle, the comforts, the wealth, the relationships, all the good things we have at our disposal improving our everyday experience? Why do these not seem to make us happier?"

Jane continued, her voice softening now that she was not championing the joys of the future. "You know, Jerry, there are a lot of very talented and wise people who have wondered the same thing. The short of it is, I simply don't know. What I suspect is that it has something to do with how evolution would select specific traits. If you pay attention to bad news, like a lion rustling in the grass who may eat you, you pass on your genes. Bad news is simply more attention grabbing, because it has survival value. The happy-go-lucky caveman who did not obsess about threats is not our ancestor. He was probably killed."

"Because bad news has such compelling power, and media outlets are designed to generate eyeballs and viewers, bad news is what we hear."

Jane continued, "We don't see headlines proclaiming, *More Humans Are Prosperous, Well Fed, Housed, and Clothed Today Than Ever Before in Human History!* "

"Here's another headline you probably won't see: *The Odds of You Dying by Violence Are a Fraction of What They Have Been in the Past!* But, according to extensive research documented by Steven Pinker in his book *The Better Angels of Our Nature*, this is also true.

"Jerry, let's table the deeper philosophical and maybe psychological question of, 'Why is nothing ever enough to help me feel satisfied?' and focus for a moment on the question about, 'Is life getting better for most people?' Don't forget the basic fact that the average life expectancy in 1900 was about 42. So, since we are both older than 42, we would probably not even be alive to ask these questions if we lived just four generations ago!"

Jane put her teacup down, wanting this conversation back on the track she had planned. The two books she had pulled out remained on the table.

"Jerry, there is good reason to believe the future is going to be amazing! Not only in giving us options we can hardly imagine right now, but in enabling emotional technologies which could mean people could be genuinely happier! What if future shrinks could figure out the mechanism and chemistry of joy, enthusiasm, bliss, and love of life? What if the crude drugs which are currently used to enhance mood could be made and individualized to enable you to feel better than you felt on your best day ever?"

"Excuse me for sounding like the college professor I was, Jerry, but I wanted to recommend to you another book which also helped me change my paradigm about the future. This is *The Rational Optimist: How Prosperity Evolves* by a geneticist by the name of Matt Ridley. It's kind of a precursor to the Diamandis book. Ridley also uses charts, documents, and hard data to show how life has improved for the vast majority of humans. It is really a fun and inspiring book, partly because Ridley reminds us of how far we have come and why we have good reasons for optimism about the future."

"Okay!" Jerry smiled as he halted what threatened to turn into a lecture from his highly educated, but sometimes didactic grandma. "You have convinced me! Or at least convinced me to read or peruse your books. If you'll leave them with me, you can trust me to read them before your next visit. I can't promise they will remove my occasional cynicism about life, however."

"I really think they may help you see a trend toward something we'll call the perfectibility of humankind, Jerry. Oh, and I promise to disinherit you if you don't return my books in good condition. But, I promised you a secret, and this whole discussion about the future has simply been to prepare you for something I probably should have shared with you years ago."

A Secret Revealed
Jane looked around the room, scooting her chair a bit closer to Jerry in a conspiratorial manner. Her voice and tone became very deliberate, her

articulation even better, as she disclosed her secret. "Jerry, because I want to see the future, and I am old, I knew I needed a way to bridge the gap that stands between where medical technology is today and where it will be in the future. Back in 1990, I signed up for cryonic suspension, which means I will be cryogenically preserved when I am pronounced legally, but not biologically, 'dead.'"

Jerry's jaw literally dropped in amazement. Here he was thinking about cryonics, and his favorite relative was already signed up! Knowing how relatively few people on earth are signed up for cryopreservation, Jerry simply found this a shockingly unlikely coincidence.

"Grandma Jane, are you serious? This is amazing! For the last few months, I have been looking into signing up for cryopreservation myself!"

As Jerry continued to ponder the shocking revelation that his forward-thinking and remarkable grandmother was signed up for cryonics, the coincidence began to diminish a bit. The willingness to try new things and the sense of excitement about the future that he shared with his grandmother probably correlated with the small but growing fraction of the population that was willing to take the gamble on cryonics.

"Jerry, this is terrific! How great would it be if cryonics could actually work, and you and I could see each other in a hundred years or so? Hey, that reminds me of another book. Thomas Friedman's recent book, *Thanks for Being Late: An Optimist's Guide to the Future,* describes how global forces are converging so that scientific advances are exponentially accelerating. The revival part could be perfected very soon!"

Jerry was fully engaged, "Wow, I can't wait to read all those books. But, meanwhile, even from what I've already learned about cryonics, I've got a big question for you. How are you handling the funding part? Cryonics is not cheap."

"Jerry, when I signed up about 30 years ago, I bought a pretty good size life insurance policy. I signed up with the Cryonics Institute in Michigan, as a lifetime member so I could eliminate any further dues. While it was a sacrifice at times to invest in the premiums, because I was already in my 60s when I bought the policy, I got as much coverage as I could afford. I thought there was a good chance the global cost of cryopreservation would go up and, like other specialty medical interventions, it has."

"I also did everything possible to stuff more money into the policy when I could, so it would have enough cash value to pay for itself. I realized that it would be hard to pay for a policy after I retired, so I paid *more* money into the policy in the early years. I don't have to worry about it now. It is guaranteed to pay the death benefit whenever I die, although I have not paid premiums in many years."

"Why didn't you tell me about this earlier, Grandma? I have about a zillion questions about this cryonics thing! Aren't you worried about all the things that could go wrong and keep you from getting a good cryopreservation? You live alone, so how would the right people be notified? What if you die in a plane crash or over the ocean? Who else have you told about this? And why the heck didn't you let me know about this earlier?"

"Slow down, dear Jerry!" Jane chuckled gently. "Yes, there are logistical questions. And there are clearly some circumstances that would prevent my body and brain from being preserved. We call these existential risks. And, Jerry, I do my best to prevent these. One of the reasons I am telling you about my cryonics arrangements is that I need your help."

As Jane said this, she straightened her posture even more in her chair in order to look at Jerry as directly as possible. Jerry found himself looking almost into the depths of her soul through those still bright eyes. "Grandson, you are absolutely right that even being signed up and fully funded for cryonics is no guarantee of getting a good cryopreservation. I need someone I can count on to help make sure that I get the best cryopreservation possible. That means if I am near death and can do so, I will go to the city where my cryonics organization is and 'die' under controlled circumstances in the hospice there."

"On this trip, I would like you to go with me. But if something happens where I can't get to my cryonics vendor while alive, I need you, Jerry, to help with the logistics of the cryo-transport team. If I can't do it, I need you to make the calls to my cryonics vendor, and maybe do some basic medical protocols. These are outlined on the wrist bracelet here." As Jane said this, she took off the bracelet on her wrist. While Jerry had noticed the bracelet before, he now understood it to be a cryonics bracelet, with instructions on who to notify, and some protocols to preserve brain pattern.

Jerry followed this line of thought and was happy to help. He was only concerned that he might not be up to the task, and he didn't want to let Grandma Jane down - especially not on something as important as the potentially life-saving technology of cryonics. And there were so many unknowns. Despite his reservations, after an appropriate amount of thinking about this commitment, Jerry responded, matching her deliberate and clear voice in making an unbreakable pledge.

"Jane, it would be my honor to do everything in my power to get you the best cryopreservation arrangements possible." He let the sentence hang in the air for a moment, wanting the level of seriousness to be conveyed to his beloved grandmother. After an appropriate moment of silence to give his pledge the dignity of a lifelong commitment, he continued. "We'll want to go over details later together, maybe even have a conference call with the cryonics organization and the cryo-transport team, but the short response is that you have my promise to be your champion to make this work."

Tears glistened in both of their eyes as these two visionary adventurers got up from their chairs to embrace. They each had a deep sense of the epic battle they were engaged in together. In the early half of the twenty-first century, two deeply sincere and well-informed individuals were endeavoring to do battle with the ultimate foe of humankind. In this room, at this moment, illuminated by Christmas lights, these people were making plans to enable the light of their lives to have a chance to continue. The drama of this epic conversation was not lost on either of them, nor did either harbor delusions that their mission would be guaranteed to succeed. In the traditions of the most noble explorers, they were setting out on a less traveled path, using the best information and technology available.

The joy they felt on this Christmas holiday exceeded any holiday they had ever experienced. To make a connection with a fellow human on this level and to be fully engaged in such a worthwhile endeavor made both Jerry and Jane feel more alive than they had felt in a very, very long time.

Chapter 7
Watercooler Wisdom

Jerry's interest in cryonics was strongly confirmed, both by finding the process was affordable and by the amazing revelation that Grandma Jane, always ahead of her time, was already a cryonics client.

And he was looking forward to sharing his research and the incredible news about Grandma Jane with his wife and best friend, Pat. Because of both of their busy schedules, the challenge was to find an optimal time for this serious discussion.

Jerry had been reading the websites of the cryonics organizations over the last week and felt pretty confident that he wanted to move forward with getting his cryonics arrangements in place.

Steve Allen was Jerry's best friend at work. Actually, his best friend period, thought Jerry to himself. Steve was single and a few years younger than Jerry. He was also a quick-witted and often opinionated counterpoint to Jerry's more conservative manner.

Right now, Steve was in Jerry's office. Well, kind of an office (actually, more like a cubicle), but its relative privacy enabled the collaborative efforts that they enjoyed working on. They worked well together as programmers, and Jerry was particularly proud that their last project had come in on time, below budget, and with very few glitches.

Testing the Waters with a Friend
During a long break at work, Jerry thought it would be a good time to get his friend's opinion on cryonics, although he found himself a bit unsure of exactly the best way to go about broaching this unusual topic.

Counseling himself to remember that "courage is only involved when you overcome fear," Jerry finally decided to take the initiative and start the conversation.

"Steve, have you ever heard of cryonic suspension?"

Steve glanced up at Jerry with an interested expression.

"Actually, it's funny you should mention that. Isn't that where they freeze people who are pronounced legally dead? I just saw a fascinating National Geographic show on that last week and, coincidentally, a show on the Science Channel just a few days ago. Yeah, it is a pretty intriguing idea, but it's really expensive, isn't it?"

"Well," said Jerry, "Maybe not. I am looking into it, and the cost can be funded with an extra life insurance policy that pays for the procedure. I am going to discuss it with my wife and I thought it would be helpful to go over the ideas with you first. You know Pat; she prides herself on being a skeptic, and any way you cut it, this is kind of a unique subject! So, I would appreciate it if I could run a few ideas by you before I talk with her. Is that okay with you?"

"Gee, Jerry, of course it is. We're best buds, right? And, you worry too much about what Pat thinks. It is your life, your money, and your decision, correct?"

"Spoken like a lifelong bachelor, Steve," Jerry observed with a good-natured smile on his face. "There is a reason I stay happily married and a bunch of our co-workers don't. I want to have my ducks in a row when I talk to Pat about this."

"These kinds of topics raise issues that are very important to people," Jerry continued, "and they go to the heart of what we believe about life. There are ideological implications here I would like to hash out with you, the same way we hash out our software engineering. I know you and I don't always agree on everything, but we share a general understanding that there is an answer, or at least a fairly optimal solution to most problems."

"We've always shared the same basic epistemology, a similar method of determining what is true. We're engineers, schooled and trained to examine a problem and find realistic and workable answers based on evidence."

Jerry continued, "What surprises me, as I discuss ideas that matter with many of the folks around here, is that some of them seem to have a different epistemology when it comes to the bigger questions in life. The idea of having a philosophy of life based on evidence instead of faith or authoritarian thinking seems to be foreign to them."

The Elephants in the Room
"You have heard me rant about this before, Steve, and I will probably rant about it again. There are two elephants in the room that everyone does their best to ignore. And they are, if you ask me, the most salient facts of our existence: *aging and death.*"

"Steve, I'd really like it if the two of us could explore cryonics as one way to rationally confront those twin elephants. In addition to the traditional religious answer to these problems, is there a way we can look at this with the same mindset we have used to tackle other difficult problems?"

Steve had been listening carefully as Jerry explained his request. There was a long pause in the conversation, making Jerry realize he had been talking a long time without a reciprocal opportunity for Steve to get a word in edgewise.

But Steve wasn't annoyed. He looked thoughtful and deliberate, which pleased Jerry because that thoughtful countenance often portended a breakthrough in their joint software development work.

"Well," said Steve, "there are some elements of this cryonics issue we can discuss rationally and perhaps come to a logical conclusion about." He raised his hand slightly to indicate that he wasn't through with this thought. "But, I also agree with you, Jerry, that there are components of this decision that are more in the arenas of religion or philosophy. Those parts may or may not yield to rational analysis. You and I know only too well that human behavior is often anything but rational. And folks even arguably smarter than you or me have spent lifetimes debating and even killing each other over matters of religion."

"However, Jer, with this caveat, I would find the discussion of this a fascinating exercise. Let's agree on a deliverable that will define a worthwhile outcome. How about a set of questions and answers that will help you deal with concerns and issues that Pat or your parents might raise? Would that be helpful?"

Jerry was once again reminded of how grateful he was to have a friend and colleague like Steve, with his insight, logical mind, and caring spirit. Steve was already breaking the problem down into bite size elements, with a deliverable. Steve also helped Jerry remember that, without carefully defined parameters, this project could devolve into a philosophic discussion that could go on forever with no conclusion.

Steve had grasped the elements of Jerry's rather loosely articulated plea for help, and Jerry knew his colleague's restless mind was already looking at the cryonics idea from various angles.

"You know, Steve, I think a written question and answer format would be a real help here. There are some 'frequently asked questions' on the cryonics websites already, but many of those seem to be more concerned with the techniques or technology of cryonics."

Jerry continued, "I'd like our work to cover the ideological components of cryonics. These are the kind of questions that are likely to come up when Pat and I talk, especially when I talk to my folks later. You know, the kind of questions that could make the next family reunion a bit awkward."

The Devil's Advocate Does His Job

"I've got some news for you, my friend," Steve smiled, but spoke with conviction. "I doubt any questions, or even answers, we can logically prepare are going to eliminate all the awkwardness that some families have at gatherings!"

Jerry was taken aback by his friend's intensity.

"Steve, I am talking about conversation with people like Pat, my wife of many happy years. I really doubt that any of them will disown me for trying to take advantage of a leading edge *medical* procedure."

"Well, Jerry, they probably won't. I am just saying that your choice in this matter could lead to some consequences and repercussions you need to be prepared for."

Jerry didn't quite know how to take his friend's serious observations and the perceived negativity that was emerging from the conversation.

"Now wait just a minute, Steve. First of all, this is not really anyone's business but my family's and mine... and I can handle family and friends, thank you. What makes you think my decision to elect to do cryonic suspension might cause such a ruckus? Unless I make a big deal out of it, how would anybody even know about my arrangements? And, furthermore, I thought you were supposed to be in my corner. Are you trying to talk me out of this idea before I even explore it?"

"I'll handle the second question first," Steve said calmly, clearly trying to defuse the annoyance he heard in Jerry's voice. "Of course I am on your side. I may consider signing up for cryonic suspension myself sometime. But you will recall, Jerry, that you asked me to help you consider all sides of this issue. And that means playing the devil's advocate at times. And I think you just need to consider that there are some people who are going to find your decision to be strange. You simply need to carefully consider the ramifications."

"So, what is your point here, Steve? That we are still living in the second dark ages where superstition is ascendant and where people don't embrace enlightenment values and critical thinking?"

"Exactly!" Steve was excited. "That is exactly my point, and it relates to how people, including people you care about, might respond to your telling them that you are signing up for cryonic suspension."

"Thanks, Steve, I appreciate your concern, but the people I care about are generally pretty reasonable folks. First of all, I don't have to tell anybody I don't want to tell. Besides, I am not asking anyone to do this besides me, although I do hope Pat will consider it over time. I have no intention of evangelizing, or proselytizing, or even sharing this with most people. It is simply no more their business than any other choice I make with my own body. Jeez Louise, Steve, this is a *medical* option, not a philosophical statement. Or as Dr. McCoy would say, 'I'm a software engineer, not a philosopher!' "

"If I needed a heart bypass to fix what nature and age have done to me, no one would give my electing to do that operation a second thought. They wouldn't assume I was trying to cheat death, do something against nature, or make a selfish or desperate decision. Of course, they would understand we do what we can to preserve our individual lives! What is the difference between signing up for cryonic suspension and undergoing a necessary, life-saving operation?"

"Jerry, I love ya Buddy, but you are, respectfully, not

thinking this through carefully. *Of course,* you will have to tell people you are close to. I would be surprised if the cryonics companies don't have family members do some kind of sign off. And, *of course,* there are ideological and even theological implications to this idea. This is an idea pretty far ahead of its time, which may be difficult for some people to accept. Let's face it, Jerry, you have always been an early adopter. Pioneers *always* have it the hardest."

Steve continued earnestly, "Even though the concept and practice of cryonics has been around for over forty years, if that documentary was right, I hear the numbers are still in the thousands, not the tens of thousands, who are signed up. I must say that I am impressed that you are so far ahead of the curve here, Jerry."

"Thank's, Steve," replied Jerry. "As I have researched, I have found that the people who *are* signed up are among the smartest and best educated people on the planet. They are the thought leaders for the emerging future. But yes, you are right, there are not a lot of them - yet. A disproportionate percentage of them are, interestingly, software engineers who are working on really cool projects like AI and robotics."

Steve admitted he was impressed by this new fact, but he couldn't let Jerry off the hook on his earlier point. "Comparing cryonics to heart bypass operations could be a bridge too far. Think about this, Jerry. Heart bypass operations are done every day, by the thousands every year. They are a proven medical therapy, widely adopted by mainstream science and the medical community. We know heart bypass operations work because people recover from them and go on to live productive and enjoyable lives. You and I work with and know lots of folks who have had heart operations. Our boss Wayne had that triple bypass a few years ago, and he recovered in a matter of weeks. Although it would have been good if they could have given him a personality transplant and a better brain while they had him in the shop!"

Steve laughed at his own joke.

"In contrast, cryonics is still virtually unknown. Most people have no idea that there is even an option that exists in the real world. The idea of cryonics is science fiction to them. What they know about cryonics they saw in the movie *Demolition Man*, or some even worse science fiction movie."

"Hey, watch it!" interjected Jerry. "I liked *Demolition Man*. But, you're right, most people are probably unaware of cryonics as a currently practiced and legitimate medical intervention."

Steve jumped in, "Even lots of folks who have seen an article or a Discovery Channel show on cryonics think it is only for very wealthy and eccentric people. Did you know, for example, Jerry, that Walt Disney is cryogenically frozen?"

"Actually, Steve, Disney is not. That is one of the myths of cryonics."

Steve wasn't going to let this correction ruin his point. He was on a roll, his large intellect fully engaged, and he was determined to get out his whole list of devil's advocate points.

"Well, okay... maybe Disney is not. But, wouldn't you agree that most people think it takes gianormous amounts of money to be cryonically suspended? And that the number of people actually signing up to do this is small? And that there is a perception that this whole thing is basically just for the uber-rich with big egos?"

"*And,* like I said before, this is far from a proven science. My point is that this is NOT comparable to a heart operation. I mean, how many people have been thawed, or unfrozen, or whatever you call it, and lived?"

"Well," responded Jerry, "Exactly zero. I am not trying to get ahead of the facts here. Reality check? As of this moment, there is not even an animal, not a cat or dog or rat or hamster or rabbit, who has been successfully resuscitated from full liquid nitrogen temperature."

"There you go," exclaimed Steve triumphantly. "You just proved my point. This is why cryonics is completely different from a heart operation. And probably why so few people have signed up. Plus, you only do it to dead people."

"Actually, heart operations, and especially heart transplants, were seriously controversial in the early years, according to an article I read in <u>Scientific American</u>," countered Jerry. "I forget the exact title or month, but it reminded me that the history of science is replete with examples of medicine interfering with nature, and the religious authorities of the day having a fit about it. And if we were to go back and study it, I bet nearly every time some scientist or researcher came up with a technology to allow humans greater control, the advance was decried by some in the religious community as 'playing God.'"

"You know, Jerry," Steve smiled warmly, "I don't always agree with you, but I love our conversations and your use of language. Although, if we are being strictly scientific, it would help if you provided more specific references to your alleged historical facts. But I forgive you for this, because exactly none of my other friends use 'replete' in normal conversation. Could you give me another example where less-than-scientific beliefs impeded human thriving?"

"Well, let's see," responded Jerry. "How about 'If we were meant to fly, we would have been given wings?' And even today, there are some groups that do not believe in or even allow their children to take advantage of modern medicine. Instead, they just pray over their sick child, asking God or gods to change the laws of nature due to their supplications. Some groups don't allow blood transfusions for similar reasons. Why? Because it is considered interfering in nature, usurping the role of the Gods, who alone should decide who lives and who dies."

"Oops… sorry, Steve; looks like I got on a bit of a rant again there. This may be a good place to stop this discussion for a bit. I don't want to get too annoying."

"Jerry, I think we've made a good start on our discussion. Let's table this for now and get back to what Wayne is paying us to do. Breaktime is dangerously close to over. This software is not going to write itself. But let's agree to continue this discussion, put some thoughts on paper, and talk tomorrow. Okay?"

Jerry smiled and patted Steve on the back as he did so. "That is indeed a deal, my friend. Thanks for being so cool about this. I think a wise man said, rather recently in fact, that this software code is not going to write itself."

Chapter 8
The Merkle Matrix

Jerry and Steve continued their discussions the next day over their lunch trays, which they carried to the conference room.

"So," began Steve, "What's new with the cryonics project?"

"Actually, I've been thinking about what a cool idea this is, now that I know about funding it through insurance," Jerry answered. "Why in the world so few folks have signed up so far remains a mystery to me. You may be right about the implications of this being different from other medical procedures."

"Let's see, we set our goal for today to begin to identify the most relevant arguments and get them down on paper."

Steve went to the white board and uncapped a dark blue marker. "Relevant? Okay, what if we start with the really relevant question of whether or not this whole thing will work? That's about as basic as it gets, don't you think?"

Jerry was ready for this. "You know, Steve, last night I was looking up "cryonics" again and came across Ralph Merkle's site. He's a well-known nanotechnology researcher and he's come up with a visual that really shocked me because it was so simple and so profound. It's a matrix about the possible outcomes of cryonics working or not working. May I show it to you?"

Steve handed the marker to Jerry, who drew a large square box and divided it into four equal sections. While he was doing this, Steve popped "Ralph Merkle" into a search engine and came up with the "Ralph Merkle Cryonics Website."

"Ok, I have the site on my laptop," boasted Steve, silently marveling at how efficient research on the web has become. "And I see the Merkle matrix you're talking about. Let me scan the rest of the site to get the context." After reading a moment, Steve continued, "Wow. This site seems to be a good introduction to our exact question, 'Will cryonics work?' answered from the point of view of a nanotechnologist. Of course, I see this guy is on the board of directors at Alcor, so I'd expect his take on cryonics to be positive."

Jerry reproduced the matrix on the whiteboard squares and they both continued reading on Ralph Merkle's cryonics site:

Overview of Ralph Merkle's Cryonics Site:
1. *The purpose of cryonics is to save lives and restore health.*
2. *Today's medical technology can't always keep us alive, let alone healthy.*
3. *A future medical technology, based on a mature nanotechnology, should be able to preserve life and restore health in all but the most extreme circumstances.*
4. *Tissue preserved at the temperature of liquid nitrogen does not deteriorate, even after centuries of storage.*
5. *Therefore, if current medical technology can't keep us alive, we can instead choose to be preserved in liquid nitrogen, with the expectation that future medical technology should be able to both reverse any cryopreservation injury and restore good health. (Merkle. Web)*

"Interesting!" said Jerry. "Check out his observation about clinical trials, which I hadn't really thought of." They continued reading:

Evaluating Cryonics:
The major reason that cryonics is not more favorably viewed in the medical community is relatively easy to explain. Medicine relies on clinical trials. Put more simply, if someone proposes a technique for saving lives, the response is, "Try it

and see if it works." Methods that have not been verified by clinical trials are called "experimental," while methods that have been tried and failed are rejected.

In keeping with this tradition, we would like to conduct clinical trials of the effectiveness of cryopreservation to determine whether it does (or does not) work. The appropriate trials can be easily described. Cryonics proposes to preserve people with today's technology in the expectation that medical technology of (say) the year 2100 will be able to cure them.

Thus, the appropriate clinical trials would be to:
- Select "N" subjects.
- Preserve them.
- Wait 100 years.
- See if the technology of 2100 can indeed revive them. (Merkle. Web)

Steve looked over at his research partner and expressed his appreciation for Merkle's insights. "Merkle makes a fascinating argument in this next paragraph. Here's our question…"

Does Cryonics Work?
The correct scientific answer to the question "Does cryonics work?" is: "The clinical trials are in progress. Come back in a century and we'll give you

a reliable answer." The relevant question for those of us who don't expect to survive that long is: **"Would I rather be in the control group, or the experimental group?"** We are forced by circumstances to answer that question without the benefit of knowing the results of the clinical trials.

In order to show that cryonics will not work (or even to show that it's unlikely to work) it is necessary to show that no future technology, no matter how advanced, will ever be able to restore the cryopreserved patient. When we consider what is routine today and how it might have been viewed in (say) the 1700's, we can begin to see how difficult it is to make a well-founded argument that future medical technology will never be able to reverse the injuries that occur during cryopreservation. (Merkle. Web)

Having put the matrix on the whiteboard, the two readers were anxious to read how Merkle analyzed his graphic tool:

What to Do:

	It works	It doesn't work
Sign up	Live	Die, lose life insurance
Do nothing	Die	Die

How might we evaluate cryonics? Broadly speaking, there are two available courses of action: (1) sign up or (2) do nothing. And there are two possible outcomes: (1) it works or (2) it doesn't. This leads to the aforementioned payoff matrix. In using such a payoff matrix to evaluate the possible outcomes, we must decide what value the different outcomes have. What value do we place on a long and healthy life? (It is important to realize that the kinds of medical technology required to restore today's cryonics patients will almost certainly be able to restore good health for an extended period). What (presumably negative) value do we place on being dead? And finally, in the absence of direct experimental results in one direction or the other, what estimate do we make of the chances that it will work?

While different people will answer these questions in different ways, this provides a useful framework in which to consider the problem. (Merkle. Web)

After a careful reading of the site, Steve looked up at Jerry. "I really like the way Merkle writes. The clarity of his thinking makes these big questions understandable. But don't you think he has somewhat oversimplified the question of whether cryonics will work? "

"What do you mean? What is oversimplified?" Jerry quizzed.

"Well, look at the Merkle Matrix." Steve pointed to the whiteboard. "Can it really be that simple? To boil a life or death decision down to a four-box matrix? There must be variables we are overlooking."

"Well, not really. Admittedly, cryonics, the medical processes, the application, and the multiple variables involved; all that isn't simple at all. But, Steve, the decision as to whether to sign up and give this technology a chance, frankly, is *exactly* that simple."

Binary Choices?
Jerry continued, "These really *are* binary choices."

"Choice One: If you don't sign up, and you die, you are dead, in all likelihood permanently; finito, not alive, not aware, unable to make choices about your existence. Similar to the Monty Python sketch about the parrot, you are an ex-person.

Choice Two: If you *do* sign up, and a technology in the future *does* revive you, you get to be alive. If you do sign up, and cryonics does not work, the worst thing that happens is that the life insurance premiums are wasted. And, I read somewhere your life insurance proceeds might be able to go to your secondary beneficiaries. So even your life insurance premiums are not wasted."

"Stevie, note that only one of the interior boxes in the matrix provides the kind of outcome we might consider optimal: signing up for cryonics and then cryonics successfully reviving you. It takes both to enable this thing to work."

Steve looked thoughtful as he spoke. "You know, I like the way Merkle explains that cryonics is indeed an experiment - an ongoing one - with the results of the experiment unknown until perhaps fifty or a hundred years from now. Merkle, like the other cryonicists I've read, is definitely not trying to oversell this."

"And," Steve continued, "I was also intrigued by the rather profound question, 'Do you want to be in the *experimental group*, or the *control group*?' I can't stop thinking about that question. And the more I think about the idea of cryonics as an experiment, with controls who do not elect cryonics protocols and with experimental subjects who do, it kind of seems like more of a reasonable gamble."

"Exactly!" exclaimed Jerry, excited that Steve could see the logic of the Merkle reasoning. "There just isn't that much downside to the risk/reward equation. You and I know there are always trade-offs every time we enable a functionality in our code work. But the tradeoff, the downside risk if you want to call it that, in cryonics is that you might waste a few bucks a day on a life insurance policy."

"No, Jerry. Once again, my friend, you are not thinking this whole package through. There is indeed more of a downside risk, as you put it, to signing up for suspension. For starters, there is more than just your cryonics life insurance premium. The cryonics organizations have membership dues that must be paid as well, if I recall the information I read on the websites."

"Good point," acknowledged Jerry. "That is indeed a factor. With the prepaid standby included, the Alcor dues are about 500 bucks a year, or $43 a month. And, to be fair, these dues could go up in the future. The membership dues at the Cryonics Institute are less at about $10 a month. So, I guess a person should add in these costs to get a more comprehensive picture of the total downside risk to this gamble."

"Jerry, besides dues, don't forget the other kinds of downside risks we talked about yesterday; like how some people could consider cryonics a 'reputation risk'."

"Jeez, Steve! There you go again; you're giving me the same arguments, and I'll give you the same answer. I think you are all wet about this reputation thing. Remember, if I want to stay 'in the closet' as a cryonicist, that is certainly my choice."

"Point taken. Meanwhile, we have to put something on this whiteboard besides the Merkle Matrix, amigo."

"Actually, Steve… not in this chapter we don't."

Steve looked at Jerry as if Jerry had just sprouted a second head from his left shoulder. "Jerry, that is a strange thing to say. What do you think we are? Characters in some book created to represent different ideas? That is just silly. Don't you feel real to yourself?"

Jerry responded, "Well, yes, I do… but then we would, wouldn't we?"

Rolling his eyes with an understanding grin, Steve observed, "Jerry my man, I have to tell you this entire line of thought is *exactly* the kind of thing that makes some of our friends think you are a bit too far out there. Don't make it easier to marginalize or dismiss your creative ideas by speculating we may be just characters in somebody's book! And besides, this is a complete digression from our task at hand, which is to provide a set of solid questions and answers about cryonics."

"Oops. Sorry, Steve, you're right. The deliverable you referred to when we first started to talk about cryonic suspension is straightforward. We still need that set of questions and answers to help explain this idea and to eventually make decisions about cryonics. Anything else is peripheral and reduces focus on our mission."

"And, tell you what, Steve, since our lunch is almost over… if you don't have big plans, could you come over on Saturday? Pat and the kids will be at her parent's this weekend. You and I could have uninterrupted time to share some serious discussion and some serious barbecue."

"I'll put it in my calendar, partner. Meanwhile, I found an interesting set of questions and answers relating to our quest, written by a different cryonicist than Ralph Merkle. I'll send you the link, and we'll see if his Q&A can help. It's Erik Krastel's website; he's a software developer. When I glanced at the site, it looks like he's done some of our homework for us already."

"Hey, Jerry, you fellow character in a hypothetical book, let's put those FAQ's in the next chapter," said Steve with a grin.

"Done deal, Steve. I'm looking forward to our discussion! See you Saturday. Back to work!"

Chapter 9
Objections Overruled

The next Saturday, Steve and Jerry sat at Jerry's backyard picnic table in front of their laptops.

Jerry started the discussion as he handed his friend a lemonade. "So glad you told me about Eric Krastel's article, 'A Defense of Cryonics.' His FAQs are amazingly on point to the questions we've been thinking about, but admittedly haven't quite gotten around to writing down!"

"You're right," Steve answered. "This article is really a great find. Thinking about Pat, which of Krastel's questions should we start with?"

"Good question. I know my wife pretty well, so we can probably leave out some of this. But I must say that all of Krastel's arguments are thought provoking."

"Your wife and I are both card carrying skeptics, Jerry. You probably know I subscribe to not one, but TWO magazines devoted to science, reason, and skepticism. So here's my vote for the first two questions to start with."

Q. I'm not going to sign up until they revive someone!

A. The desire to see someone revived before signing up for cryonics is very common and, on the surface, seems to make sense, but the irony is that once the ability to revive cryonauts is developed, there'll be little need to suspend people, since the entire point of cryonics is to put you "on pause" until that technology arrives. (Krastel. Web)

Q. Once you're dead, you're dead! How can you come back?

A. How do you define death? Please be specific.

The definition of legal death is not a concrete demarcation, but rather changes as medical technology advances. By that definition, people have been "brought back from the dead" numerous times.

Unless you are obliterated by some violent action, life and death are not binary states. Death is a cascade failure of biological systems, usually starting with cardiac and/or respiratory arrest and ending with ischemia, apoptosis, and decay. In between is a gray area. I think that a good analogy is a line of dominos falling. The objective of cryonics is to halt or suspend that cascade until such time as the damage that has already occurred can be repaired. (Krastel. Web)

"Hmm," said Jerry. "These do seem like foundational questions. And I understand and appreciate skepticism in general. But, you're so skeptical, you remind me of the guy who refused to put wood on his fire until it gave him some warmth! Here, I think Pat would resonate with these next three."

Q. How do you know that people in the future will bother to try to revive you?
A. There are several reasons why: desire to prove that it can be done, to revive family or friends that are in suspension, to advance science, or general altruism. If you're looking for a completely economic/rational reason, as technology continues to improve and spread, eventually the cost of reviving patients will be less than the cost of maintaining their stasis. (Krastel. Web)

Q. What happens to your soul?
A. I really can't answer that question for you, but I can let you know that cryonics doesn't violate the tenets of any major religion (except maybe Scientology, if you consider that a major religion). It's a medical procedure, like an organ transplant or open-heart surgery. It's not "raising the dead." (Krastel. Web)

Jerry spoke up because he had been thinking about how to respond to this exact question about souls.

"Hmm, interesting way to think about this. Krastel's argument makes sense. When Pat's dad was revived during a heart operation, after they thought he was dead, no one worried about this soul question. If we look at this as a medical intervention, the parallel is obvious."

Q. All your friends and family will be dead! Won't you be lonely?
A. They won't be dead if I can help it! I'll hardly be alone. My wife is an Alcor member, as are three of our close friends, and I expect that number to grow.

Additionally, the cryonics community is small and fairly close-knit. I've had the pleasure of meeting quite a few Alcor members and staff, and without exception, they're all pleasant, interesting people that I'd be happy to have over for game night or spend a few decades next to, insensate in an steel dewar. I like these people, and we'll all be going through this adventure together. So, if nothing else, we'll have each other to lean on.

Even setting all that aside, people make new friends in new environments all the time. (Krastel. Web)

Steve looked over at his friend. "Jerry, just like the aspirational goal of our software projects, this guy is 'all killer and no filler!' These are exactly the kinds of questions most reasonable people might ask. And in these next few Q&A's he confronts some of the more pragmatic issues."

Q. You won't have any usable skills in the future!
A. We all start out life with no marketable skills. I learned everything I need to function in society once, and I'm perfectly willing to do it again. Besides, even a job as the future equivalent of a fast food worker or sewer repairman would be infinitely more appealing to me than the alternative. (Krastel. Web)

Q. The culture shock in the future will be terrible!
A. Yes, it probably will be! I expect it to be disorienting, frustrating, and at times scary, and possibly even depressing. This isn't for everyone, and I don't mean to portray it otherwise; however, the situation will probably not be unlike that of an immigrant from a third world nation coming to the US and making their way. It's not easy, but people do it every day. Furthermore, cryonicists are a self-selected group, and I doubt that any of us have signed up without giving it a lot of consideration first. The people who sign up for cryonics (in my experience) are fascinated by the future, in love with life, and are the most likely to view this as a worthwhile adventure, even in light of the potential downsides. In other words, it's a group that self-

selects people who are likely to be psychologically predisposed toward being able to deal with the inevitable stresses. (Krastel. Web)

Q. Doesn't it cost a fortune?
A. Fortunately, not. I'm not wealthy by any means, and I know cryonicists who are on much shakier financial ground than I am. The cryo suspension itself is typically paid for using a life insurance policy. If you're in good health and are reasonably young, then a policy is very affordable. (I can recommend an insurance agent who is familiar with setting up such policies.) The insurance policy builds interest, and after a while (20-25 years or so), the interest will likely be sufficient to actually make the payments for me.

In addition to the insurance policy, there are quarterly dues to Alcor itself. The total dues are currently $130 per quarter, with substantial discounts for students, additional family members, and children under 18. For this, you're getting the basic membership, tags, subscription to Cryonics magazine, and (this is important) access to Comprehensive Member Standby, which basically means that, in the event that you are at high risk of legal death in the short term, they'll send a team to be ready and make preparations beforehand. (Krastel. Web)

After reading the cost section out loud, Jerry reminded Steve that other cryonics organizations had different financing structures.

"Yeah, I remember reading the global costs at the Cryonics Institute were lower than Alcor's, but I didn't analyze the possible tradeoffs," Steve responded. "Hey, as a skeptic, I really like these next slightly more technical explanations."

Q. What happens if the power goes out?
A. Not much. Contrary to popular belief, the dewars that patients are stored in don't use electricity at all. Instead, they're kept cold using liquid nitrogen. The only outside intervention they need is to have the nitrogen topped off from time to time (it naturally boils off slowly). Liquid nitrogen is cheap, safe, and easily obtained from multiple sources (it's actually a waste byproduct of some industrial processes).

As for redundancy, there are multiple sources for liquid nitrogen, so even if the supplier they contract with vanished overnight, there would be no emergency. For that matter, even if every supplier vanished, they could likely go a few months before the situation became critical. The nitrogen boils off from top to bottom (obviously), and full-body patients are stored inverted so that if every source was somehow simultaneously cut off for an extended period of time (which would indicate some serious problems with the world as a whole), the head (the most critical component) would be the last to thaw. (Krastel. Web)

Q. Won't ice formation cause irreparable damage to cells? Won't your brain turn to mush when it's thawed?
A. Not if there's no ice. The whole point of perfusing the cells with cryoprotectant prior to suspension is to eliminate the water inside and between the cells and replace it with a medium that won't cause as much damage when it crystallizes. Even that process has now been upgraded to a better one.

A few years ago, Alcor made a significant change to their neurosuspension (head) process when they switched from "freezing" to a process known as vitrification. The important difference is that with vitrification, ideally there'll be no crystal formation at all. The vitrifying solution never actually "freezes." Instead, it goes through what's called "glass transition." The result is an unparalleled suspension.

Within the last year, Alcor developed the ability to do vitrification for whole body patients as well, *though the technology still has room for improvement.*

As a side note, Michael Shermer, a gentleman for whom I personally have tremendous respect, based his critique of cryonics almost entirely on damage from ice crystal formation. (Krastel. Web)

Q. What makes you think that vitrification is better, or will work at all?
A. Because vitrification has been reversed for individual organs, and those organs have been perfectly viable. This is with current technology. Granted, the process has only been reversed for certain organs and not for entire organisms, but it shows that this is a workable process, and that cryonics' goals are really only an evolutionary step beyond what's currently capable. There's nothing magical required here, and no violations of the laws of physics. (Krastel. Web)

Jerry took a moment to observe, "These technical notes are helpful. Back to your skepticism, however, I am impressed that Krastel doesn't shy away from responding to popular perceptions and misperceptions about cryonics. Just take a look at how he deals with these criticisms."

*Q. Didn't Penn & Teller say that cryonics is bull****?*

*A. Yup, they sure did. I'm a huge fan of P&T and was really excited to hear what they had to say on the subject. Unfortunately, they really only breezed over it. They gave a quick briefing of what it is and declared it bull**** without providing any real supporting arguments. Instead, they showcased a guy who runs a very small cryonics operation (TransTime). The guy was obviously nervous on camera and gave a poor interview. Their segment makes no mention that there are any other cryonics organizations (there are at least 4 others), despite the fact that they use footage of Alcor's facility, and despite the fact that Alcor has suspended more patients than every other cryonics organization in the world put together.*

They did interview one medical professional whose only argument against cryonics is that ice crystal formation damages proteins. (Krastel. Web)

Q. Won't you just come back old and feeble, and die again shortly thereafter?

A. Er, no. Aging is largely a function of degenerative cellular damage. It's likely that reviving a patient in cryo suspension will require the ability to repair damage on a cellular level. In other words, aging will likely be conquered before most people are brought out of suspension. (Krastel. Web)

Q. Wasn't Walt Disney frozen?
A. This isn't an objection, but I've been asked this several times, so I figured I'd include it. No. And yes, I'm sure of this fact. Walt Disney died on Dec. 15, 1966. The first real cryonic suspension was Dr. James Bedford on Jan. 12, 1967. He is currently located at Alcor's facility in Scottsdale, AZ. (There's an interesting story there, but it's beyond the scope of this article.) Walt Disney is in a grave. (Krastel. Web)

Q. Who would win in a fight between ninjas and dinosaurs?
A. Ninjas, of course. Why do you think that the dinosaurs are extinct? Okay, I admit, nobody has ever asked me this. I thought I'd break things up a bit. :-) (Krastel. Web)

The two men had a good laugh at the ninja reference. Jerry noted, "I like this guy's sense of humor. Kind of helpful, since we are talking about life and death! But this Q&A coming up is my favorite of all, since I am such a fan of Ray Kurzweil and the incredible impact he has had on the world."

Q. Would you still be human if you were uploaded or had a synthetic body?
A. "Will we still be human? In my view, this is exactly what being human is all about: expanding

beyond our limitations." -- Ray Kurzweil (Krastel. Web)

Steve looked at his friend. "Seems like our work has been done for us! Thank you, Eric Krastel! Hey, Jerry, remember that 'It's better to emulate genius than to create mediocrity!' Let's print this out, and I vote we name it 'The Official Deliverable!' "

"Now, about the barbecue and beer you promised me..."

Chapter 10
Happy Wife, Happy Life

"In sure and certain hope of the resurrection, we commit the body of your servant, Terry Grossman, to the earth."

The priest finished and looked up at the small group of people gathered at the cemetery. The afternoon was hot, and the priest, like everyone else, was getting uncomfortable. He addressed the group, "Your presence here today was very important to the family. Thank you all for coming."

Jerry and Pat headed back to their car, greeting friends along the way, sharing remembrances of the well-beloved Terry Grossman.

Remembering a Friend
As Jerry opened the car door for Pat, he was deep in thought about life and death, existence and nonexistence, and the profound implications of these last few hours.

As they pulled onto the main road, Pat broke the silence. "I know you don't like funerals, Jerry, but I appreciate your having come. Terry died so young and had so much left to offer! It seems such a waste!"

Jerry responded with great sincerity, "You are so right, Pat. Terry was a truly unique and irreplaceable part of so many people's lives, including ours. I've been thinking that every death is kind of like a library burning down. Terry's particular brand of wisdom, forged by his unique experiences, now seems so lost to the world. I am increasingly resistant to the idea that we are powerless against death. I know, 'death and taxes' and all that... but you'd think that with all the progress that we humans have made, we'd be able to do better."

"I'm thinking along the same lines, Jerry. Yeah, Terry's death... so young and so close to us... makes even *me* want to know more about this cryonics thing you and Steve have been looking into. Based on what you've seen so far, what do you think?"

"Well, since we have some time in the car here, and because we both just looked the reality of death in the face, I'd be happy to share what I've learned so far. The bottom line is that I'm pretty intrigued. There are no guarantees, of course, but even the possibility that, at some point in time, a mind like Terry's could thrive again instead of being irretrievably gone, is huge."

"I can't think of a better time for us to talk about this than right now. Let's go over some of your thoughts on cryonics, Jerry," responded Pat.

Beginning the Cryonics Conversation

"Okay," he began, "I want to do a decent job explaining this, and I don't pretend to be an expert, because I am not. I'm a software engineer, not a cryobiologist or scientific researcher. Basically, cryonics is a medical procedure that is kind of an ambulance ride to the future. It's currently only performed on folks who are pronounced legally dead. But, apparently, being legally dead does not mean your brain pattern immediately goes away. If you are cooled down quickly enough, the memories and patterns that make you '*you*' may be preserved for resuscitation at some point in the future."

"I love the idea, Jerry. But, as a reality check, how credible is this? Is it legitimate science? People have been trying to cheat death since caveman days, and we're pretty sure it hasn't worked yet. What makes us think this cryonics thing might be different?"

"Honestly, Pat, at this point, even the experts can't unequivocally demonstrate the viability of cryonics. I'd love to say that a laboratory somewhere has revived a dog from full liquid nitrogen temperatures, but I don't think that's happened yet."

"But here's the thing… Studying the way science and technology are having exponential growth curves, it's easy to see the arc of progress. That's what kind of convinced me that this is a credible concept. Can I tell you my personal favorite three arguments?"

"Okay," Pat agreed.

"First, we know that there are thousands of people walking around today who were at one time just sperm and eggs in liquid nitrogen."

"And, we know that occasionally people who were pronounced 'dead' on operating tables are later revived, with, amazingly, full brain functions intact.

And, you've probably heard that people, especially children, have fallen through the ice and been underwater for surprisingly long periods and later were revived."

"Oh, wait, Pat! Here's another point. I've found out that some current heart operations are performed at reduced temperatures to increase the time that circulation can be stopped without permanent brain damage."

"Wow, Jerry. You have been doing some reading and research on this, haven't you? I always knew my husband was smart. Let me stop you long enough to ask a bottom line question. Isn't this cryonics thing just for really rich people?"

"Pat, when we get home, I've got some FAQs Steve and I printed out from our research that may be on point to some basic questions. And thanks for the compliment. I think you're smart, too! And you and I both want to know about the cost piece. Do I hear you suggesting that you'd like your smart husband to be an *affordable* immortal?"

"That may or may not be funny, Jerry. But right now, I need to get some approximate idea about costs. Why even bother to spend any time talking about this if it's not affordable for us?"

The Affordable Immortal
"I've been looking into the cost," Jerry responded. "It turns out that most people fund their suspension with an extra life insurance policy. So, while the actual cost of the suspension may be $200,000 or more, that cost is actually born by a life insurance policy that costs as little as a few bucks a day."

"Oh. Well, that may be okay. I just don't want you spending a bunch of money on a cryonics scheme and leaving me and the kids high and dry when you go."

"Pat, you know I would never do anything like that. I'd *never* make a decision like this, any more than you would, without both of us feeling good about it. That's why we have our current personal life insurance policy, plus the extra $500,000 I have at work which goes to you if something happens to me."

HONK HONK!

The car behind Jerry was honking. Jerry looked up to see the light was green; perhaps it had been so for some time. "This is why one should not multitask on important issues," he thought to himself with some chagrin. But this conversation with Pat was very important to him.

"Well, I'll tell you what, Jerry," Pat was finally responding. "Let's continue to think about this idea, and you keep doing your homework on it. I love you and trust you and respect your judgement. I just don't have any opinions yet because I don't have enough information. Like everything else we do, we'll get smart - find out all we can - and then we'll make the decision together."

As a comfortable silence fell in the car again, Jerry reflected how everything about this day had heightened his appreciation for the value of each relationship, for the value of each person, and for the value of life.

Chapter 11
Any Excuse for a Vacation!

Jerry and his wife both had professional jobs and they shared housekeeping chores. This included the ritual of emptying the dishwasher and stacking the dishes in the cupboards, which they were doing now.

The discussion about cryonics had been in play for several weeks now, since the crucial conversation following their friend Terry's funeral. As he had promised, Jerry had continued to do his homework.

As an initial way of finding out cost factors, Jerry had filled out a Quote Request Form he found at the rudihoffman.com website. This enabled him to get price quotes for various amounts of life insurance.

Additionally, his many questions on life insurance and cryonics had been answered through several phone and video conferences with Rudi Hoffman. Jerry liked the fact that he could schedule these one-on-one conferences himself, at times convenient to him, through the online scheduling calendar available on the Hoffman website. He especially appreciated the fact that these conversations were without cost or obligation.

Jerry had also communicated with staff members at Alcor and at the Cryonics Institute.

"Jerry, I've been considering this cryonics thing and doing a bit of reading online about it myself," Pat said, as she handed another plate to Jerry. "I still don't believe we have all the pertinent information necessary to make an optimal decision. What would you think of going to one of the actual places where they do this and witnessing it personally, and see how we feel then? The Cryonics Institute is in Michigan, and Alcor is near Phoenix, right?"

"You are right about the locations, and that is a great idea, Pat!" enthused Jerry. "You know, you and I have a vacation coming up, and the grandparents have already volunteered to stay with the kids. Alcor would make sense for us since it is so much closer, although they are both good organizations. What if we fly to Phoenix, do the home office tour, and then continue to Hawaii for the rest of our vacation? I think we could have almost a week in Hawaii, and I am pretty sure this whole package can still be affordable."

"Hmm, Jerry, I do like the idea of actually seeing what Alcor looks like, meeting the people, and feeling what kind of vibes the operation has. Before we make the cryonics decision, it seems to me it's worth the trip to know as much as we can. Honey, I'm thinking we're going to Alcor and then on to Hawaii! Aloha!"

Chapter 12
Mecca for Nerds
(A Visit to the Time Machine)

The plane ride had been crowded, but still fun, thought Jerry as they emerged into the Phoenix airport. He and Pat had sat down with their budget and figured out how to go to Hawaii by way of Phoenix. "Pat," Jerry said with a grin, "I know we have mentioned this before, but since we live near San Francisco, Phoenix was not exactly on the way to Hawaii!"

Pat was in a playful mood, as she had been ever since they started planning the trip to Alcor and then Hawaii. "Well, maybe it is not on the way, but if we hadn't actually planned this and put this trip together, there is no convenient time for it to happen, Jerry. I have heard you say a million times, 'Success is not convenient.' "

The logistics of getting their bags and getting an Uber were remarkably seamless. Jerry felt this was a good omen for the Alcor tour and also to what he was thinking of as a second honeymoon for Pat and him. My goodness, he needed this vacation!

Phoenix seemed to be a very large, sprawling town, with sections spreading into the Arizona desert. Sherry, their Uber driver, told them that they were going to Scottsdale, an upscale section of Phoenix, and pointed out items of interest on their ride. It was clearly hot a good part of the year in Phoenix, and since it was a safe, but enjoyable topic, the conversation centered on the weather. "It was over 114 degrees a good part of last summer," said Sherry. "And you guys are going to Alcor, which I guess is a cryogenics lab where they store people at very cold temperatures? Seems like they chose a curiously hot place to have their lab!" Sherry smiled pleasantly.

Jerry reflected on his many hours of cryonics research and felt rather proud to actually know why Alcor was located here. "Alcor used to be in California, but because of the constant threat of earthquakes, they moved to Arizona, which is much more geologically stable. Nobody really knows how long the patients will need to be in liquid nitrogen, and they don't want to lose them in an earthquake or tsunami."

"Oh, I guess that makes sense," noted Sherry as she wheeled the Prius into a pristine area of commercial and industrial buildings. "We are going to 7895 East Acoma Dr. Unit 110, in Scottsdale. Should be right up here."

Jerry and Pat were looking around, curious about the layout and location of Alcor. The heat was starting to shimmer off the pavement. It had been pleasant earlier, but the day was going to get seriously hot.

"And there it is!" shouted Jerry triumphantly, as Sherry parked. Jerry and Pat got out; he hit an additional tip on his smartphone for Sherry, retrieved their luggage, and looked at Alcor from the outside.

Visiting Alcor
"Looks clean and professional to me. What do you think, Pat?" he said as he wheeled the suitcases toward the front glass door. "I read that Alcor owns the whole building in this commercial area and does a leaseback for the part they currently use. They have also continued to improve the physical plant, including security."

This security was becoming obvious because the front door was locked. An attractive young woman let them in through the glass door, greeting them warmly. "I am Diane Cremeens, the director of membership. We spoke several times on the phone. You must be Jerry? And this is your wife Pat? Welcome to Alcor. If you put your luggage in this closet, it will be safe, and we can start your tour."

The front room and the entire Alcor facility was organized, welcoming, and professional. The overall impression was that of a high-end medical facility. As they were shown around and introduced to the staff and leadership (about a dozen paid staff and volunteers), the solidity and integrity of the entire operation was evident.

The highlight of the tour for Jerry and Pat was seeing the actual cryonics dewars, which were about two meters across and four meters tall. "We used to let people go into the dewar area," explained Diane, "but for security reasons, we now have this large area with bulletproof glass that allows us to view the patient section."

Both Jerry and Pat gazed with awestruck wonder at the dewars, gleaming stainless-steel containers. There were dozens of them stretching the length of a room the size of a basketball court. Each one was actually a container within a container, with a hard vacuum between the layers of stainless steel to reduce heat transfer and maintain a constant temperature of - 196 Centigrade. In the containers were over one hundred actual human beings, pronounced "dead" by current medical standards, waiting on the march of human progress to change the standards.

The couple were impressed by the cleanliness and orderliness of the entire facility. And, when they were sharing a soda in the executive boardroom, several of the staff members they had met on their tour came in and provided further details on how they fit into the "Alcor family."

"I would love to meet the CEO, Max More, if he is available," Jerry mentioned as he and Pat selected a seat at the conference table. "I have read many of his writings and articles. He and his wife Natasha Vita-More are real thought leaders in an important philosophical movement called Transhumanism, and it would be great to meet them in person."

"I'm certain Max would be delighted to meet you both as well," responded Diane, "but he and Natasha are speaking at different conferences this week. Max is at a conference in England, meeting with cryonicists in the UK to improve our ability to handle the logistics of cryo-transport from Europe, and Natasha is speaking at a conference in China."

"Many, but not all, of the folks who work here are personally signed up for cryopreservation," Diane explained. "And some of the people who are here as volunteers or staff have been involved for many decades. Speaking of that, may I introduce you to one of the co-founders of Alcor? Coming in the door right there is Linda Chamberlain. Linda and her husband Fred started Alcor as a registered non-profit in 1972. Now Fred is cryopreserved here, while Linda helps with special projects. This year, Linda and several of us on staff are going through the paperwork of the entire membership, making sure the life insurance or other funding arrangements are solid."

"It is a genuine pleasure to meet you, Linda!" Jerry was delighted and surprised at the same time. After introducing his wife to Linda, he asked, "Linda, could you sit and join us for a moment? I am truly happy to meet you. I have been reading over the past several months about the history of the cryonics movement. And, here, live and in person, and looking remarkably energetic and healthy, is one of the founders of Alcor!"

Linda Chamberlain was petite and attractive. She seemed a bit uncomfortable as Jerry was going into full "groupie mode" in meeting her. As Jerry and Pat talked with Linda, Diane excused herself to go back to her office, not neglecting to give them hugs and inviting them to say goodbye before they left Alcor.

Taking advantage of the opportunity to hear the history and development of Alcor directly from Linda, the two visitors delved into the history, controversies, medical and technological issues, and most importantly, the financial structuring of

Alcor. It was a long and detailed discussion, covering some of the same ground Jerry had covered with Pat in her earlier skepticism.

Linda displayed remarkable depth of knowledge about nearly every component of the organization. What impressed both Jerry and Pat the most, however, was that Linda was completely candid when a question could not be answered. The three of them spent a lot of time on the logistics of how cryonics members could get good cryo-preservations, and Linda went into some detail about the local hospice that partners with Alcor.

Jerry was thrilled that Pat was getting to hear this information directly, since he knew that there was no way he could have properly transmitted the nuances to her secondhand. There are just some subtle components in the communications process that need to be done in old fashioned "meatspace," he thought to himself.

The key issue here was one of trust. And it was clear that Pat was having a great time, fully engaged, asking both general and technical questions. The conversation was sprinkled with laughter, as Linda shared some of the more bizarre but interesting bits of cryonics history from the point of view of someone who had lived that history.

Then it was time to leave. As they poked their heads into the offices of the people they had met earlier at Alcor, it was clear that this visit had been a strong confirmation of what Jerry had decided some time ago. He could trust these people to do the right thing.

Heading to the hotel in a different Uber, this one captained by a rather quiet, older gentleman, Jerry and Pat reviewed their impressions of their Alcor tour.

Later, after checking in and getting back down to the pool, they enjoyed a refreshing dip. As they lay side by side in the surprisingly comfortable loungers, the arid air of Arizona seemed to magically dry their bodies faster than expected.

"I'll tell you what, Pat," started Jerry, "Let me bring us some Pina Coladas from the bar, and let's see if we can come to some sort of determination on Alcor."

"Hey, big spender, that drink will cost ten bucks here. How about we split one?" responded Pat.

"Practical as always, my pet," said Jerry as he made his way to the bar.

Sharing the even more expensive than expected Pina Colada, Jerry and Pat sat back in their loungers and took a moment to just breathe and enjoy the beauty around them.

Jerry turned to Pat, glad they were in the shade from the still-bright late afternoon sun. "Pat, I am really glad we are taking this trip. Do you think we could have these next eight days together and really not worry about anything else except loving each other?"

"Wow," said Pat wistfully. "That is a charming thought, but the reality is that we are probably going to be worrying about the kids and everything else we have to think about, including the costs and logistics of this trip. With work, the kids, our home life, and wondering about taking care of my mom and dad, we have a lot on our plates. But we can certainly try to relax. And I really thought you wanted to talk about this cryonics thing."

And Now, Decision Time

"So much for dreaming! You're right, Pat. I am eager to get your thoughts on cryonics. And could we have a lovelier place to talk?"

"I enjoyed our visit to Alcor, Jerr. The people involved are truly well-meaning, competent, and professional. The tour of the facility, the technical details they provided, and the operating room demonstrated that to me. It's obvious that they are far from delusional about cryonics. But that still doesn't mean the science, especially the revival part, is quite ready for prime time yet."

Jerry responded, "You're right. And, interestingly, nobody pretended it was. If anything, Pat, skeptics like you and me become more suspicious rather than less if we encounter a "true believer" mentality. And this did not feel like a high-pressure sales situation. Did you get any crazy cult vibes from the Alcor experience?"

"Definitely not. In fact, Jerry, if anything the vibe was anti-cult-like. Most of the people working at Alcor are signed up for cryonics themselves, but they were not required to be, like I might have expected. I actually found that rather confirming. They aren't making promises on things they can't guarantee. And I love their candor about what is real now and what is still speculative."

Pat continued, "In short, Jerry, now that we know more, and you have found out what the costs will be, let's go through all the pros and cons again."

Over the next few hours and a lovely dinner, Jerry and Pat revisited every issue they could think of regarding the cryonics decision. After covering the medical, technical, and philosophical issues, they proceeded to discuss the cost component.

Jerry agreed, "Yep, let's do this one step at a time. How about we figure the total if we sign just *me* up for now, and then the global costs if we *both* sign up?"

"Sounds like a plan. Use that brilliant, nerdy, mathematical brain of yours and let's make sure we're dealing with the exact costs. The global cost you keep talking about is a combination of membership dues plus the life insurance premiums, right? And in your individual situation, this is exactly how much?"

Jerry's Costs

"Well, Pat, aren't you clever to know that I know?" laughed Jerry. "Rudi gave me quotes on several different coverage amounts. The recommended ideal coverage is $400,000 of the Index Universal Life on me. My premium, based on my age and health, is $356 a month. The membership dues paid monthly are another $43 a month, bringing the total to $399 a month. A happy husband for under $400 a month. My raise this year is more than that." The corners of his eyes crinkled as he winked at her. "How is that for a deal?"

"You could get signed up with a lower face amount, though, right, which would, of course, be a lower investment?" Pat asked. "How does that work? Let's say you have a policy for $400,000, and right now the process only costs $200,000. Who gets the other $200,000?"

"Good news, Pat," explained Jerry triumphantly. "You do! Remember the extra coverage is for a cushion to handle future cost increases at Alcor. The difference between whatever Alcor charges at the time the service is needed and the $400,000 goes to you and the kids. And don't forget the $500,000 group term insurance I have from work also goes to you."

"We also talked about maybe adding a cryonics trust later on. The cryonics trust is a separate option that may enable some of our money to grow for us while we are in suspension."

Pat jumped in. "Whoa. We said we weren't going to get into details about a possible cryonics trust right now. But I do understand that a cryonics trust is another reason to have *more* life insurance coverage rather than less. I recall an attorney you mentioned who does these trusts. Peggy Hoyt? We may want to work with her later. But for now, let's stay focused on our main cryonics decision."

"Pat, remember, with the Index Universal Life, a good part of the money goes into a cash value; that's the part of the policy you don't have to die to get. And, since I do indeed have a mathematical brain, I even remember that the projected cash value on my policy at age 100 is over $1.3 million!"

"Got it," responded Pat. "Because these policies grow cash, the money we reposition into them can be budgeted as a savings plan as well as life insurance."

"Jerry, my dear, you spent several hours boring me with your newfound insights into the way modern policies work. I have to admit the way money can grow in these policies is really phenomenal and, I admit, I wound up being pretty amazed! Here's how I understand it: in Index Universal Life, cash value grows inside the policy at good rates since it is indirectly tied to the S and P 500. That cash value can be withdrawn tax-free any time (subject to some constraints). So, it's kind of like a Roth IRA, since you can pull the cash out of the ending values, tax-free."

"And in your policy illustration, Jerry, the one done specifically for you, the cash value at age 100 grows to over one million dollars? That is a crazy high number! This life insurance policy we're looking at is set up to pay a tax-free death benefit of $400,000 to cover your cryopreservation."

She continued to summarize, "Since Alcor does not need this full amount right now, we'll have the extra coverage go to me, tax-free. The part that goes to Alcor is also tax-free. And on top of that, the policy grows a *cash value* that we can pull out as a living benefit on a tax-free basis."

All of a sudden, Pat started laughing. Jerry looked at her, somewhat confused, as the matters of finance and life insurance didn't strike him as the most amusing and light-hearted of topics.

Now she was practically giddy with something she had come up with!

"Jerry, what if Benjamin Franklin was wrong? What if you *can* beat death *and* taxes?" Pat was overly delighted with her own observation.

"I like it!" exclaimed Jerry. "Sounds like a good subtitle for a book! What if we actually live in a time when a confluence of technologies might let us beat both death and taxes? What would old Ben Franklin have done? I bet he would have been among the first to sign up for cryonics!" Jerry observed gleefully.

It gave him a warm feeling of confirmation, thinking that there were clearly visionary humans who saw past current limitations earlier than others, and that he was in the process of joining those pioneers.

Pat had finally stopped laughing, and they both knew this inside joke would be one they would treasure and enjoy together in the future.

"Enough about you, dear. Tell me what my costs look like."

Pat's Costs

"Well, remember the good news is that since you're younger and female, your rates are even better. $400,000 of the Index UL of coverage on you is only $254 a month. Yes, we could start with lower face amounts, and add more coverage later. But then we'd have to jump through the underwriting hoops again to prove insurability, and we'd pay the premium based on our age at that time. So, if we can handle it, I think it's in our long-term best interest to lock our current age and good health in now."

"Alcor dues are only $30 a month for a second family member. So the global costs are pretty doable. Recall that our car payment ends this month, and that will reduce our outlay by $500 monthly. And since you'll now be insured by this policy to protect the boys and me, we can save another $150 by consolidating your old policy into this higher yielding one."

Pat spoke deliberately, "You know, I'm liking this. I think I want to do it. To have a reasonable chance to continue our adventures in a brand-new world is just too good to pass up. I'd love to be in the future with you. I still have some unanswered questions we'll want to talk about, but I'm feeling like we're doing the right thing."

"Jerry, as soon as we get home, let's go over that *Ten Step Cryonics Checklist* and call Rudi. He can email us the life insurance applications and then help us contact Alcor. For now, we have a second honeymoon in Hawaii to enjoy!"

Time to Celebrate Life!
The couple looked out at one of the most beautiful sunsets they had ever seen. They weren't used to being able to see such a broad panorama. The sky was ablaze with shades of red, green, blue, brilliant orange, and vivid yellow. On the eastern horizon, stars were beginning to emerge. The clear, dry air of the desert displayed a universe of new worlds, each one a potential future destination. The possibilities of their many tomorrows beckoned.

Jerry looked at the love of his life, recognizing their life-changing pact. Pat smiled back at him warmly. He took her hand in his. He breathed deeply, inhaling the clear air and feeling more alive than he had in years. Life was good. He had reached deep within himself to execute a project that had a realistic chance to enable life to be even longer and better for both him and for his life companion. Jerry experienced a genuine thrill at the wonder of being alive, as he spoke with a quiet intensity to his wife.

"Tomorrow is going to be a beautiful day, for the rest of our lives and beyond…"

Appendices
(Stuff you probably ought to know that did not fit seamlessly into the book narrative.)

In going through multiple iterations of this book over the last 17 (!) years, I have endeavored to keep the version you have read (and hopefully enjoyed!) on point and as succinct as possible. I trust you have learned something and that I was successful in making the arcana of both cryonics and life insurance interesting.

I care deeply that any individual who is considering cryonics has convenient access to all the critical information they require. The following appendices are included to inform your decisions about cryonics and funding choices.

Appendix 1
Five Takeaways

Here are five of the big themes that I hope were obvious in this book and that you take away from your reading.

1. Cryonics is a legitimate, though currently unproven, medical intervention.
2. You can choose to be in the cryonics "experimental group" and not in the "control group" for this particular long-term experiment.
3. Cryonics may be affordable for you through the leverage of life insurance. (You did get this, right?)
4. If cryonics does indeed work and you are revived, it will probably be in a really spectacular and fun future. (Wouldn't it be cool for us to hang out together in the future?)
5. There are resources and people to help you in your research and decision making. I am one of those people.

Appendix 2
A Ten-Step Checklist for Becoming a Fully Signed and Funded Cryonicist

- [ ] 1. Go to www.rudihoffman.com and fill out the "Quote Request" form.

- [ ] 2. Set a phone/skype video visit appointment on the web-based calendar.

- [ ] 3. View the four short videos under the "Cryonics" tab of the website.

- [ ] 4. Discuss options with Rudi Hoffman; make informed decisions on cryonics vendor, amount and type of life insurance preferred.

- [ ] 5. Sign the pre-completed application sent to you; mail or scan/email this back to Rudi.

- [ ] 6. Complete the local nurse exam and separate health history phone call.

- [ ] 7. About 6 weeks of underwriting later, you receive your policy. Return any needed requirements to put the policy in place.

- [ ] 8. Contact the cryonics organization and complete their application. (Rudi will have already sent them a full copy of your policy.)

❏ 9. Receive your cryonics bracelet and/or neck chain.

❏ 10. Congratulations, you deserve to celebrate. You are a fully signed and funded cryonicist!

Appendix 3
A Note About Life Insurance Ownership and Cryonics Organization Requirements

Amazing reminder about life insurance:
It avoids probate and taxes!

Life insurance proceeds go *directly* to a named beneficiary by operation of law. This means, as a practical matter, that it is virtually impossible for anyone to interfere with or supersede your plans, since there is no probate process involved.

In other words, your life insurance proceeds avoid probate, and do *not* require you to have a will or trust to make absolutely certain that the face value will go to the cryonics organization for your funding payment. Cool, huh?

Important note on cryonics policy ownership:
Understandably, both Alcor and the Cryonics Institute require that your financial arrangements *guarantee* that they get paid.

The practical impact of this requirement is that, generally, the cryonics organization is named both as a *direct beneficiary* (of all or a part of the proceeds) *and* as an *owner/joint owner* of the policy.

Should this worry you?
No, in fact it should be confirming to know that there are solid mechanisms in place to protect your cryonics organization from major unreimbursed financial liability.

This does *not* mean that your cryonics organization can access your cash value or change your policy without your consent. Nor does it mean you can't change your mind, redirect the policy proceeds, and get all your cash value. It *does* mean that the cryonics organization will know about any substantive changes and be required to sign off on any major changes to the policy.

Most carriers disallow corporate ownership.
Since most widely recognized life insurance carriers don't permit corporate ownership, cryonics organizations work only with a limited number of insurance carriers. By the way, this wasn't always the case, but it is now.

Fortunately, there are some excellent, highly rated, cost effective, over-engineered and consumer-oriented carriers who ARE more than happy to write these "Cryonics Friendly" policies.

These are the carriers I work with and have developed deep and mutually beneficial relationships with over decades. These are the carriers I personally own policies with, and who I actually trust my life and my clients' lives with.

Contacts and References

Contacts:

Rudi Hoffman CFP(r)
www.rudihoffman.com
386-235-7834
rudi@rudihoffman.com

Alcor
www.alcor.org
7895 E. Acoma Dr. #110
Scottsdale, AZ 85260
877-462-5267

Cryonics Institute
http://www.cryonics.org/
24355 Sorrentino Ct.
Clinton Township, MI 48035
586-791-5961

Peggy Hoyt
The Law Offices of Hoyt and Bryan
www.HoytBryan.com
407-977-8080
(Peggy and her colleagues are the world's leaders in executing cryonics trusts and are adept at working at distance and/or with your attorney.)

Books:

There are a tremendous number of great books available on transhumanism, philosophy, science, technology, agnosticism, and general self-improvement. I have about 900 books on these topics and related topics running around my brain (that occasionally works as intended). Personal note: audio/audible books have dramatically improved and changed my life. Here are a few of my favorites.

De Grey, Aubrey. *Ending Aging: The Rejuvenation Breakthroughs That Could End Human Aging in Our Lifetime.* New York: St. Martins Press. 2007. (Aubrey is a personal friend-- and hero to me and most anyone else following anti-aging and age reversal science. With his trademark Methuselah beard, Aubrey has become the recognized leader in the charge to end the tsunami of involuntary death. He has created SENS (Strategies for Engineered Negligible Senescence) as an umbrella organization for this purpose. Aubrey is also an "out of the closet" cryonicist. SENS is worth checking out and contributing to, at *www.sens.org*.)

De Wolf, Aschwin and Stephen Bridge. *Preserving Minds, Saving Lives: The Best Cryonics Writings from the Alcor Life Extension Foundation.* Available at Alcor. (Anthology of great writing by the history making thought leaders of Alcor, curated by my good friends Aschwin and Stephen, who remain cryonics activists.)

Diamandis, Peter. *Abundance: Why the Future Is Better Than You Think.* New York: Free Press. 2012. (Amazing book by a larger than life guy who has started 15 companies. Along with Ray Kurzweil, a predictor of the future - largely because they are helping to create it.)

Diamandis, Peter. *Bold: How to Go Big, Create Wealth, and Impact the World.* New York: Free Press. 2015.

Drexler, K. Eric. *Engines of Creation: The Coming Era of Nanotechnology.* New York: Anchor Books, 1986. (One of the first books on nanotech, and the amazing possibilities and disruptions possible through Molecular Nanotech. A classic, and responsible for many cryonics signups. I still remember the thrill I felt reading this book.)

Drexler, K. Eric. *Radical Abundance: How a Revolution in Nanotechnology WIll Change Civilization.* New York: Public Affairs. 2013.
(27 years after *Engines*, Drexler hits it out of the park again with the possibilities, as well as the dangers, of molecular manufacturing and medicine.)

Ettinger, Robert. *The Prospect of Immortality.* New York: Doubleday. 1964. (Kind of the book that started cryonics. I treasure my personalized copy. Ettinger was a college professor by trade, but his real contributions are his seminal work about cryonics and the future. Unflaggingly optimistic, a bit naive about the financial piece of cryonics, Ettinger remains a hero to many of us involved in cryonics.)

Friedman, Thomas L. *Thank You for Being Late: An Optimist's Guide to Thriving in the Age of Accelerations.* New York: Macmillan Publishing Group. 2016. (Well researched and fun to read. Documents how fast things are changing, especially since 2007, according to Friedman, the year that a number of curves went pretty much straight up. No wonder we feel we can't keep up. We can't, but not to worry, it will all be for the best.)

Istvan, Zoltan. *The Transhumanist Wager.* Futurity Imagine Media. 2013. (A powerful, watershed book. I have curious and contradictory thoughts about this book; it is both one of the best and one of the worst books I have ever read. Make up your own mind, but this one will make you question some paradigms.)

Kahneman, Daniel. *Thinking, Fast and Slow.* New York: Farner, Straus, and Giroux. 2011. (Brilliant book about cognitive biases and the various experiments used to identify them. I listened to the audiobook twice and wanted to see his charts, so I bought the hard copy.)

Kurzweil, Ray. *The Singularity Is Near: When Humans Transcend Biology.* New York: Penguin. 2005. (Any of Kurzweil's books, *The Age of Intelligent Machines; The Age of Spiritual Machines: When Computers Exceed Human Intelligence;* or *Fantastic Voyage: Live Long Enough to Live Forever* will blow your mind in a very positive and potentially life changing way. I have been following Ray's career and reading his books since the 1990's. He is the Thomas Edison of our age and the track record on his predictions is remarkably accurate. I understand he is signed up for cryonics at Alcor, and I wish he'd be more "out of the closet" about this.)

Pinker, Steven. *The Better Angels of Our Nature: Why Violence Has Declined.* New York: Viking. 2011. (Pretty much what you'd think from the title. Because he understands how much pushback he'll get to this idea, his documentation is exhaustive and meticulous. Pinker's use of the language is about as precise as a human can be.)

Pinker, Steven. *Enlightenment Now: The Case for Reason, Science, Humanism, and Progress.* New

York: Penguin Random House. 2018. (I just now ordered this in hardback from Amazon, based on reading articles, excerpts, and a compelling speech a client forwarded me of Pinker on YouTube. Evidently a lot of people missed the memo about Humanism and the Enlightenment and want to take us back to the bad old days of the Dark Ages, so Pinker has to explain that science is good, and superstition is bad.)

Ridley, Matt. *The Rational Optimist: How Prosperity Evolves.* New York: Harper Collins. 2010. (A truly wonderful book and audiobook. Ridley is a geneticist by training, but this book is about human progress. Has there ever been a better chapter title than the first in this book, "When Ideas Have Sex" ?)

Websites Cited:

Merkle, Ralph. *http://www.merkle.com/cryo/* continually being updated by Merkle. (You may know Merkle is, in addition to the creator of the Merkle Matrix, the inventor of "Merkle Trees" that are foundational in the math of blockchain/bitcoin technology in ways I simply don't understand.)

Krastel, Eric. "A Defense of Cryonics."
https://www.xeromag.com/alcor.html

(Long term friend, client, cryonics activist, and wonderful human.)

Urban, Tim. "Why Cryonics Makes Sense." *https://waitbutwhy.com/2016/03/cryonics.html* (Probably the best article ever written summarizing points about... well, "Why Cryonics Makes Sense.")

Acknowledgements

This book almost did not happen. Fortunately, I have in my life people who encouraged me. The hard drive of my computer has some ten different iterations of this book, generated over the last 17 years. I really hope you like this iteration and find it helpful.

To my patient wife Dawn, my partner in both life and business: my deepest love and gratitude. I am glad you and I have made and properly funded cryonics plans to be in love for a long time. You are my inspiration and my muse. I love the life we have made over the last 34 years together. My apologies for not taking some of your suggestions about the book earlier. If I had, maybe it would not have taken 17 years.

To my editor, mentor, and professional author coach, Sean Donovan: I am most sincerely glad that we met. You put up with more angst and weirdness than most relationships would bear, and you have stayed helpful and professional throughout our months of working together to make this book a reality.

To my sister Trudi Taylor: you are a continuing inspiration to me and you are always a steadfast and loyal supporter of my worthwhile endeavors, including this book. The surprising amount of time we invested in continual improvement of this book has been cherished and appreciated. No one could ask for a more wonderful friend and sibling, and I love you.

And to the pioneers of the anti-aging, age reversal, and cryonics movements: my undying respect for your vision and possibility thinking. Special thanks to Eric Krastel and Ralph Merkle for permission to use their brilliant writing.

And to you, dear reader, for your taking a piece of your valuable life to read this book.

About the Author

Rudi Hoffman is a fellow human being, sharing about 99.9% of the same DNA you have. He works hard at being a genuinely good human and has a passion for sharing ideas that matter with his fellow travelers on the planet.

He lives in the beautiful village of Port Orange, Florida, with his wife Dawn, administrator for Rudi Hoffman CFP(r), and with their three spoiled dogs. Professionally, he maintains the top three credentials in financial planning, CFP®, CLU, and ChFC. Licensed since 1978, he maintains life insurance licenses in 49 states and in the District of Columbia. He is in the top 1% of life insurance brokers worldwide, having enabled hundreds of millions of life insurance benefits to be available to wonderful and grateful clients.

Rudi is the world's leading authority on funding and financial matters relating to cryopreservation. He is overly proud that more than 70% of the funded cryonicists on Earth have been assisted in their cryopreservation financing by his firm. He is also humbled by the responsibility this represents. He would be pleased and honored to have you view some website videos, fill out a website form and schedule an appointment to talk with him.

Rudi can be easily reached at:
Email: rudi@rudihoffman.com
Phone: 386-235-7834
Website with "Quote Request" form and short videos under the "Cryonics" tab
rudihoffman.com

Made in the USA
Monee, IL
23 March 2021